Springer Series in Advanced Manufacturing

Series Editor

Duc Truong Pham, University of Birmingham, Birmingham, UK

The **Springer Series in Advanced Manufacturing** includes advanced textbooks, research monographs, edited works and conference proceedings covering all major subjects in the field of advanced manufacturing.
The following is a non-exclusive list of subjects relevant to the series:

1. Manufacturing processes and operations (material processing; assembly; test and inspection; packaging and shipping).
2. Manufacturing product and process design (product design; product data management; product development; manufacturing system planning).
3. Enterprise management (product life cycle management; production planning and control; quality management).

Emphasis will be placed on novel material of topical interest (for example, books on nanomanufacturing) as well as new treatments of more traditional areas.

As advanced manufacturing usually involves extensive use of information and communication technology (ICT), books dealing with advanced ICT tools for advanced manufacturing are also of interest to the Series.

Springer and Professor Pham welcome book ideas from authors. Potential authors who wish to submit a book proposal should contact Anthony Doyle, Executive Editor, Springer, e-mail: anthony.doyle@springer.com.

More information about this series at http://www.springer.com/series/7113

Josip Stjepandić · Markus Sommer ·
Berend Denkena
Editors

DigiTwin: An Approach for Production Process Optimization in a Built Environment

Springer

Editors
Josip Stjepandić
PROSTEP AG
Darmstadt, Germany

Markus Sommer
isb innovative software businesses GmbH
Friedrichshafen, Germany

Berend Denkena
Institute of Production Engineering and Machine Tools
Leibnitz University of Hannover
Garbsen, Niedersachsen, Germany

ISSN 1860-5168 ISSN 2196-1735 (electronic)
Springer Series in Advanced Manufacturing
ISBN 978-3-030-77538-4 ISBN 978-3-030-77539-1 (eBook)
https://doi.org/10.1007/978-3-030-77539-1

This Springer imprint is published by the registered company Springer Nature Switzerland AG
The registered company address is: Gewerbestrasse 11, 6330 Cham, Switzerland

Acknowledgment

SPONSORED BY THE

Federal Ministry
of Education
and Research

This research and development project was funded by the German Federal Ministry of Education and Research (BMBF) within the "SME—Innovative: Research for Production" Funding Action (funding number 02K18K018) and implemented by the Project Management Agency Karlsruhe (PTKA). The authors are responsible for the content of this publication.

Used Trademarks

Dassault Systemes SolidWorks:	Dassault Systemes
DYMOLA:	Dassault Systemes
Google Glass:	Google
MATLAB:	The Mathworks
Microsoft HoloLens:	Microsoft
Siemens Plant Simulation:	Siemens

Contents

Editors and Contributors

About the Editors

Dr. Josip Stjepandić is the head of business unit 3D Product Creation at PROSTEP AG, the leading product data integration company worldwide. After receiving his grade as a M. Eng. from the University of Zagreb and his Ph.D. from the University of Technology of Graz, he has worked for two automotive suppliers in the areas of engineering simulation and design methodology. From 1994 to 1996, he was an Associate Professor for Applied Informatics in Mechanical Engineering at the University of Applied Sciences Dortmund. Since 1996, he has been in charge for consultancy and solution development at PROSTEP AG in the areas of design methodology, supplier integration, systems engineering, knowledge-based engineering, product data validation and visualization, configuration management, and CAD data exchange for many industries (automotive, aerospace, shipbuilding, and machinery). One of his most important contributions is the multiply awarded global supplier portal OpenDESC.com. His current research is focused on systems engineering, modular design, and Digital Twin. He is the Vice President of ISTE, the International Society of Transdisciplinary Engineering (www.intsoctransde.org).

Markus Sommer is the CEO at isb innovative software businesses GmbH, the leading service provider in Digital Twins and behavior models with AI methods. He studied Electronics and Technical Informatics at the RWU Hochschule Ravensburg-Weingarten University of Applied Sciences. At 1997, he generated the first AI model (MLP Network) to detect a hand in static fields. In the past, he has worked for different automotive suppliers as project manager, team manager, and head of software development. Since 2017, he works as CEO for isb innovative software businesses GmbH, with focus on new technologies (www.ki-modelle.de) like AI and evolutionary algorithm in the areas of 3D reconstruction, optimization, recommendation, und configuration systems for the industries such as automotive, logistics, space, and defense.

Berend Denkena, born in 1959, is a trained machinist and studied mechanical engineering at the University of Hanover. He worked for Thyssen in Germany and the USA before joining Gildemeister Drehmaschinen in Bielefeld in 1996, where he was the head of development and design. Since 2001, Denkena has headed the Institute for Production Engineering and Machine Tools (IFW) at the Production Technology Centre of Leibniz University Hannover. Here, around 80 research assistants work in the fields of machining processes, machine tools, production organisation, and CFRP lightweight construction. Berend Denkena is the spokesperson for the Collaborative Research Centre 653 "Intelligent Components in the Life Cycle" and deputy spokesperson for the Collaborative Research Centre 871 "Regeneration of Complex Capital Goods". He also serves on various supervisory and advisory boards. He is a member of the International Academy of Production Engineering (CIRP) and the German Academy of Science and Engineering (acatech) among others.

Contributors

Sergej Bondar PROSTEP AG, Darmstadt, Germany

Berend Denkena Institut Für Fertigungstechnik Und Werkzeugmaschinen, Leibniz Universität Hannover, Garbsen, Germany

Marc-André Dittrich Institut Für Fertigungstechnik Und Werkzeugmaschinen, Leibniz Universität Hannover, Garbsen, Germany

Wjatscheslaw Korol PROSTEP AG, Darmstadt, Germany

Gina Vibora Münch Institut Für Fertigungstechnik Und Werkzeugmaschinen, Leibniz Universität Hannover, Garbsen, Germany

Klaus Seiffert isb innovative software businesses GmbH, Friedrichshafen, Germany

Moritz von Soden Bornemann Gewindetechnik GmbH & Co. KG, Delligsen, Germany

Markus Sommer isb innovative software businesses GmbH, Friedrichshafen, Germany

Josip Stjepandić PROSTEP AG, Darmstadt, Germany

Sebastian Stobrawa Institut Für Fertigungstechnik Und Werkzeugmaschinen, Leibniz Universität Hannover, Garbsen, Germany

Chapter 1
Introduction to the Book

Josip Stjepandić, Markus Sommer, Sebastian Stobrawa, and Berend Denkena

1.1 Origins of the Digital Twin

The first notion of a Digital Twin can be traced back to Michael Grieves of the University of Michigan, who introduced the topic of Digital Twin in his lecture on Product Lifecycle Management in 2003 [1]. In the early years, the concept of Digital Twin was primarily used in the NASA's Apollo project to simulate the behavior of the capsule in space. However, a project with a similar high-fidelity simulation approach, concluded by USA Air Force Research Center during the same period of time, has carried a different label: SORCER (Service-ORiented Computing Envi-Ronment) [2]. In the same period, the terms Concurrent Engineering [3], Product Lifecycle Management [4] and Systems Engineering [5] have attracted more attention in academia and industry.

The idea of the Digital Twin is to generate, maintain and use the virtual version of a system (product, production system, instance) during the entire product lifecycle (design, test, manufacture, exploitation). The aim of this holistic approach is to

J. Stjepandić (✉)
PROSTEP AG, Dolivostr. 11, 64293 Darmstadt, Germany
e-mail: josip.stjepandic@prostep.com

M. Sommer
isb innovative software businesses GmbH, Otto-Lilienthal-Strasse 2, 88046 Friedrichshafen, Germany
e-mail: sommer@isb-fn.de

S. Stobrawa · B. Denkena
Institut Für Fertigungstechnik Und Werkzeugmaschinen, Leibniz Universität Hannover, An der Universität 2, 30823 Garbsen, Germany
e-mail: stobrawa@ifw.uni-hannover.de

B. Denkena
e-mail: denkena@ifw.uni-hannover.de

J. Stjepandić et al. (eds.), *DigiTwin: An Approach for Production Process Optimization in a Built Environment*, Springer Series in Advanced Manufacturing,
https://doi.org/10.1007/978-3-030-77539-1_1

conclude a high-fidelity product definition, verification, and validation including the modes of failure before the physical system is actually manufactured. The outcome of a Digital Twin is not only the product improvement of the physical product by failure prevention in exploitation, but also better quality, higher performance, cost reduction, shortening of time-to-market, and higher customer satisfaction [1].

In the literature, a few definitions of the Digital Twin are available. This will be further elaborated in Sect. 3.1. These fairly divergent definitions are caused by the different views based on specific characteristics of various use cases where the Digital Twin is in question (industry, phase in product lifecycle, environment, technology). The meaning and understanding of the term Digital Twin have continuously evolved in the course of its advancement without a significant consolidation. The inconsistent use of the term Digital Twin makes it much more challenging to distinguish the singular expressions and identify the main components of a Digital Twin. Therefore, that conceptual blurriness spawned numerous adjacent terms, such as the Digital Shadow or the Digital Angel, which hinders precise usage of the term Digital Twin [6]. A suitable taxonomy (see Sect. 3.2) will help to understand specific differences but, of course, cannot cover all expressions of Digital Twin which become subject for Configuration Lifecycle Management (CLM) [7].

While the terminology has evolved over time, the basic concept of the Digital Twin model has remained fairly stable from its inception. It is based on the idea that a digital counterpart of a physical system could be generated as an entity on its own. This digital information would be a "twin" of the information that was embedded within the physical system itself and be linked with that physical system through the entire lifecycle of the system. The common features of the available concepts and definitions include the integration of various data sources, which allow a digital representation of the physical object or process over its entire lifecycle [1].

With the emergence of the basic IT technologies like cloud computing, Internet-of-Things (IoT), Big Data, and Machine Learning (ML), substantial developments occurred in specific industrial domains such as Industry 4.0. From the early beginning, Industry 4.0 has facilitated digital achievements as the pillars of its development [8]. Due to fact that all data are available in digital format, and sensors are inbuilt into the industrial spaces, the main pre-requisites for Digital Twin in area of manufacturing are fulfilled. A mirroring or twinning of systems between what existed in real space to what existed in virtual space and vice versa becomes possible and meaningful. Using the emerging extensive simulation capabilities, it became feasible to perform realistic tests in a virtual environment. Furthermore, the overwhelming rise of artificial intelligence, primarily expressed by techniques such as ML, made a significant contribution to the further progress of the Digital Twin [9]. Following to these technical advancements, manufacturing industry has broadly implemented Digital Twins in the past few years for specific tasks, for example, the monitoring and the predictive maintenance of long-lasting goods in service [10].

Nowadays, Digital Twin is used both as a subject of scientific research as well as a buzzword to attract existing commercial products and services [11]. Nevertheless, the implementation of the Digital Twin is still considered undertaking a complex procedure because of the vast divergence of technical and organizational options. This is also the reason why the integration of a Digital Twin with PLM is not completed yet, although the largest vendors of PLM claim their leadership also in Digital Twin [12]. Specific types and expressions of the Digital Twin are presented in Sects. 3.3 and 3.4. Indeed, appropriate reference models and architectures are subject of research, no industrial or international standard is known yet [13, 14].

The progress and development of a Digital Twin is not limited to the area of engineering only. Coming from the medical technology, the concept of Digital Twin has entered also the area of medicine in order to predict emergence of disease (for example: rise of a tumor) and develop the protecting and healing measures [15]. In a similar manner, a Digital Twin is deployed to simulate the societal processes [16]. In area of economy, a Digital Twin of economy can help to predict the impact of macroeconomic measures, for example, changes in tax policy [17].

Studying a complex system, however, cannot be fully successful when context is left out of scope. Interaction between a system and its context (for example: environment or workforce) needs to be part of the study. Transdisciplinary systems pose a specific challenge because they add a level of analysis, which does not exist on the level of each of the disciplines [18, 19]. The content of this book considers the working environment, but not the human factors.

Starting from the practice that it took nearly a decade for a new concept to be implemented, the editors and authors undertook a pragmatic approach to develop during a two-year project a specific Digital Twin offering primarily for small and medium enterprises which as service can be implemented within weeks.

In this chapter, the introduction in this book is presented: origins of the book (Sect. 1.2), goals of the book (Sect. 1.3), expected audience (Sect. 1.4), content of the book (Sect. 1.5), and finally, contributors of the book (Sect. 1.6).

1.2 Origins of the Book

This book explores the results of the research project "DigiTwin—Efficient Generation of a Digital Twin in the Manufacturing" conducted by four partners supported by the German Federal Ministry of Education and Research (BMBF) within the framework concept "KMU innovativ" in period 2018–2020. In the communication with the funding agency, this project was evaluated as the particularly successful and worth for further dissemination [20]. Therefore, it was decided to present its outcome to a broader audience. The editors were representatives of their research organizations. The authors were members of the project team.

1.3 Goals of the Book

This book is an attempt to present the latest developments and best practices of the principles of Digital Twin based on the specific experience of the editors. The presentation includes not only current Digital Twin processes and methods, but also, very importantly, a specific real-life application and concept for its commercialization. These applications and experiences are aimed to show that the Digital Twin is still an indispensable part of research and business in manufacturing industry nowadays. The term Digital Twin expresses a plethora of concepts and approaches that can be classified as Digital Twin subtypes or derivates. Each such approach must be assigned to an innovative product, process, or service development. Each approach must also consist of methods and tools to enable and support extensive collaboration and information exchange between people from different disciplines, functions, departments, or companies.

The first goal of the book is to describe the state-of-the-art, summarizing Digital Twin achievements in area of manufacturing. A second goal of the book is to illustrate the choices that exist in organizing information flow in the modern factory. These choices encompass selection of methods and tools; technical as well as organizational. The methods and tools show the variety of problems that need to be tackled in practice and can be understood as a constituent of a customer-oriented whole. They should support trade-offs and finding (near-)optimal solutions. The third goal of this book is to demonstrate that Digital Twin has become indispensable, used widely in many industries and that the same basic engineering principles can be applied to new, emerging fields like self-optimizing factory. The final goal of the book is to provide practical example and use case that thoroughly illustrate achievements and practices of a Digital Twin. This is conducted by presentation of the research project solution which helps to generate the Digital Twin of manufacturing in the built environment provided by a middle-sized enterprise. In addition, many remaining trends challenges in research and practice have been listed.

1.4 Audience

The authors intend this book to be useful for several audiences: industry experts, managers, enterprise architects, business planners, postgraduate and Ph.D. students, researchers, and software developers. The content is intended to serve both as an introduction to development and assessment of novel approaches and techniques of Digital Twin and as a compact reference for more experienced experts. In this role practitioners can use the content to improve their core competencies and use it as a reference during their daily work, in particular for planning of commercialization activities. Graduate and undergraduate students who have already mastered several basic areas of engineering may find useful instruction material to practices in modern industrial product creation processes. Researchers can find recent achievements and challenges in various fields of Digital Twin [3].

Engineers in various design domains, such as mechanical, electrical, computer science, and environmental and logistics engineering may find this book helpful to understand the fundamental background as captured in modern product, process, and service development. It may help them to understand the multi-disciplinary, multi-dimensional and multi-level nature of a Digital Twin. It may help them to request information they need from and to supply information needed for engineering processes to the relevant stakeholders. It may help to exploit the benefits of new IT techniques like contactless object acquisition or ML. It will help stakeholders from various domains to understand how a Digital Twin works and to participate in different Digital Twin teams [3].

Managers need to understand information representing numerous facets of Digital Twin for developing a comprehensive strategy and establishing suitable engineering structures and organization. The decisions they make must advance business competition by meeting quality, cost, and time targets. This book paves the way to a pragmatic solution to introduce Digital Twin. Management and engineering need to exchange information rapidly and seamlessly so that the processes will be adjusted to support the business strategy and so that management can understand and track product issues and maturity. This book presents several methods for organising, transferring, tracking, and tracing of information [3].

Students and researchers in the wide area of manufacturing need comprehensive information on recent achievements and on directions for future research. The book fulfills this need. For this purpose valuable information can be found in the closing part of this book. Each chapter comprises an exhaustive list of recent references in related area which either were used by authors or are recommended for further consideration [3].

Finally, a further audience may consist of developers of tools and development platforms who usually have a strong software engineering background and are not experienced in applications and process development. In particular for those who define and implement integration scenarios like enterprise architects, this book could be a useful reference [3].

1.5 Content of the Book

The book has been structured into eleven consecutive chapters, which will be introduced below. Singular chapters contribute to various aspects of basic concepts, methods, technologies, industrial applications, and current challenges of Digital Twin. This chapter is the introduction to this Book as explained above.

In Chap. 2 it is explored how the production planning and control can be supported by Digital Factory methods to optimize production processes and workflows. The Digital Factory has already been recognized as a strategic advantage by the industry more than ten years ago. An essential component of the Digital Factory are digital models of products, processes, and resources, which aim the standardization of processes, the increasing of flexibility and shortening of the lead time. Especially for digital models in production, there are already existing many fields today (e. g. planning of factories, optimization of layout of the shop floor, approval processes in reconstruction and fire protection, or optimization of production processes). Assuming the gap between the real and the virtual factory, concrete solutions are needed for the continuous synchronization of the planning data and simulation models from the product development, the production planning and the production by using measured values from the real factory. High quality CAD data of all geometrical objects in all stages of planning process are the pre-requisite for seamless downstream processes which must be, apart of other non-geometrical characteristics of the production system, acquired if 3D planning data are not existent or not accurate [21]. A central component of the Digital Factory is simulation, which is used to represent real objects and processes in a virtual environment. This virtual environment is suitable for performing analyses and planning processes so that understanding about the real system can be gained. Accordingly, planning processes such as factory planning, investment planning, capacity planning, bottleneck analyses, inventory planning, and internal material transport can benefit from simulation by gaining more valid and far-reaching insights. However, simulations must be designed for specific use cases in order to be able to process them. Therefore, the corresponding parameters for the use of the simulation must be determined. The approach presented here focuses on several use cases to create a framework that is not only valid for a single use case, but allows for an arbitrary application which is as comprehensive as possible. This leads to a Digital Twin, which, in turn, can handle several use cases and is not focused on one use case like the simulation. Chapter 2 deals with the mentioned applications, primarily focusing on the requirements of the use cases for the simulation framework by identifying and specifying the required parameters. Accordingly, a comprehensive list of parameters and their exact properties is presented to support production planning and control. With this understanding, an efficient determination of these parameters can be carried out in the further course of this book, from which the generation of a Digital Twin is enabled.

In Chap. 3, the conceptual view for Digital Twin is drawn as a new key approach in the field of PLM. Briefly, a Digital Twin is a digital representation of an active unique product or unique product-service-system with its selected characteristics within dedicated lifecycle phases. This concept has experienced a tremendous impact by IoT technology, which has drastically reduced the costs. It builds the foundation not only for connected products and services but also for entirely new offerings and business models. For a better understanding, a taxonomy from recent literature is presented which consists of eight dimensions and 18 characteristics at all. Three main characteristics of the Digital Twin were identified: representation of a physical system, bidirectional data exchange, and the connection along the entire lifecycle. For a better understanding, three subtypes of Digital Twin are presented, namely: The Digital Master, the Digital Manufacturing Twin, and the Digital Instance Twin which refer to the different phases of the product lifecycle: design, production, and operation. Therefore, Chap. 3 formulates a consistent and detailed definition of Digital

Twins and gives insight in dedicated research direction. Finally, based on the aforementioned Digital Twin characteristics, an approach for generation of Digital Twin in manufacturing in a built environment is shown which is implemented and described in the following chapters of this book.

In Chap. 4, techniques and approaches for acquisition of objects in a production plant are presented. A three-dimensional digital representation, as the Digital Twin of machines and objects within production plants, is becoming increasingly important for efficient planning and documentation of production. During the acquisition of the images, the two areas of accuracy and detail are the key. The accuracy describes the precision in which the Digital Twin represents the real objects. For example, it discovers how well the dimensions of a machine within a reconstructed model matches its counterpart in reality. Detail refers to the level of detail in the digital model: Should the door of a machine tool be distinguishable from the machine or not. Two basic technologies have been established for scanning. The photogrammetry method creates a dense point cloud based on images using a sophisticated toolchain. The laser scan method firstly measures an accurate model of the environment using a laser scanner. In a second step, the color information is assigned to the measured points via the evaluation of photos. For certain applications, models resulting in highest accuracy and detail are not necessary; in these cases, speed and practicability of the acquisition is the primary concern. Due to recent advancements in photogrammetry, these methods are best suited in the case of generating a Digital Twin for production facilities. Furthermore, these methods provide a better physical handling and accessibility in case of occlusion and larger object height. An overview of the methods available as well as the underlying principles are presented. Practical considerations and examples show the feasibility and results of different photogrammetry approaches, resulting in the presentation of a photogrammetric system well suited to the task of creating a Digital Twin of a production facility.

In Chap. 5, the methods and approaches of the AI in manufacturing are introduced. Nowadays, AI becomes one of the core drivers of the industrial development and a critical factor in promoting the integration of emerging technologies, in the new generation of Big Data and Industry 4.0. In particular, ML as a particular expression of AI becomes increasingly more frequently applicable in manufacturing applications. Chapter 5 presents a systematic overview of today's applications of ML techniques and approaches for their usage in the manufacturing environment. The utilization of ML methods is explored related to manufacturing process planning and control, predictive maintenance, quality control, in-situ process monitoring, control and optimization, logistics, robotics, assistance, and learning systems for shopfloor employees. Exhaustive fundamental and problem concept describes how to select an appropriate ML approach including the concatenation of multiple approaches. Supervised methods dominate the state of the art with reinforcement learning methods gaining even more attention in recent years. Subsequently, the gains of ML such as derivation of model configuration based on the data, generation of behavioral models through training, easy validation of the model and optimization of the real system based on the model are illustrated. This illustration is surrounded by two use cases from plant engineering. Data analysis points out how these concepts can be analyzed and with behavior model can be achieved regarding the implemented ML method. Finally, conclusions and outlook reflect the benefits and drawbacks of ML and draw the way for future development.

In Chap. 6, the methodical approach for object recognition based on a point cloud is presented. Recognition of an object from a point cloud, image, or video is an important task in computer vision which plays a crucial role in many real-world applications. The challenges involved in object recognition, aiming at locating object instances from a large number of predefined categories in collections (images, video or, model library), are multi-model, multi-pose, complicated background, occlusion, and depth variations. In the past few years numerous methods were developed to tackle these challenges and have reported remarkable progress for 3D objects. However, suitable methods of object recognition are needed to achieve added value in built environment. Suitable acquisition methods are also necessary to compensate the impact of darkness, dirt, and occlusion. This chapter provides a comprehensive overview of the recent advances in 3D object recognition of indoor objects using Convolutional Neural Networks (CNN). Methodology for object recognition, approaches for point cloud generation, and test bases are presented. The comparison of main recognition methods based on methods of geometric shape descriptor and supervised learning and their strengths and weakness are also included. The focus lies on the specific requirements and constrains in an industrial environment like tight assembly, light, dirt, occlusion, or incomplete data sets. Finally, a recommendation for use of existing CNN framework for implementation of an automatic object recognition procedure is shown.

In Chap. 7, aspects related to data quality in the complex processes are discussed. The challenge of enhancing and generalizing interoperability as an important prerequisite for Digital Twin is often hindered by the fact that data quality requisites depend on the purpose for which the data will be used and on the subjectivity of the data consumer. Data quality is getting as important as product quality in manufacturing process. In Chap. 7 it will be presented how to systematically handle the data quality requirements and support a comprehensive data quality management. After the definition of Digital Thread as a data highway, the classification of data quality is discussed based on data quality dimensions and standards related to field of manufacturing. The data quality metrics is discovered upon the guidelines developed by national and international harmonization bodies from global automotive industry. Three practical examples from design and manufacturing as well as data migration in industrial context give insight in practical challenges and achievements in the field of data quality as well as future directives. The discussion section emphasizes the high importance of data quality for the generation of a Digital Twin as a seamless process chain.

In Chap. 8 practical implementation of the object recognition based on methods described in Chaps. 4–7 is explored. In order to prepare the point cloud for a feature-based process of planning and simulation, the objects in it must first be recognized and a CAD-based structure must be created. The great challenge in every process chain is to implement the desired function with minimal effort and losses. As one of the alternatives, the object recognition is selected to implement the layout planning of a factory in 3D. In Chap. 8, it is demonstrated how to embed the object recognition in the process of virtual 3D layout planning in a built environment as well as which findings and results can be expected. The aim of Chap. 8 is to investigate and evaluate the usefulness of a realistic 3D virtual factory model in factory layout planning primarily for the Digital Twin based on object recognition. This is addressed by a practical study of how existing methods for data acquisition and processing can be concatenated and, subsequently, applied under real industrial constraints and conditions. During this study, realistic 3D layout models are created using point clouds acquired by commercial terrestrial laser scanner and prepared for object recognition with CNN considering the strict data quality requirements and further constraints. The selection of models was discussed and the results were evaluated in industrial workshops with engineers involved in the layout planning and machine operators that will work within the production system. Seamless, robust, (semi-)automatic workflow of primarily standard, modular components with low user assistance was of particular interest. Chapter 8 is concluded with the discussion of the achieved results, the solution alternatives, and the present approaches how to extend the utilization, improve, and simplify the entire process.

In Chap. 9, the design of simulation models based on previously provided CAD models is explained. Setting up a virtual model of a production system is time-consuming and requires expertise in the use of simulation software as well as knowledge of the operations in production systems. For this reason, the creation of virtual models as a basis for the Digital Twin of a production system is hardly feasible for many companies. Chapter 9 presents an approach for the efficient generation of simulation models. For this purpose, production systems are first systematically described using an ontology. On the one hand, this description serves to develop a generic creation method. This is necessary to be able to convert any production system into a virtual representation. On the other hand, the ontology provides a structured data model for subsequent modelling of the Digital Twin. In Chap. 9, an approach for generating a simulation model of the shop floor is introduced, including its functional behaviour as well as a visual representation. Since most steps of the approach are automated, the simulation model can be generated highly efficient. Also, less expertise in handling simulation programmes is needed. The remaining manual steps require either no special IT knowledge or can be carried out via services.

In Chap. 10, the planned commercialization of the research project solution is shown. Product and service innovations are the key driver of the competitiveness in the manufacturing industry. For long time, partnerships between manufacturers, suppliers, and service providers have been established to improve overall performance of entire supply chain. Most members in such a supply chain try to expand their business by developing platforms and building their ecosystems. This also occurs in the IT-related industries, in particular when services are provided for large companies. In Chap. 10, the commercialization of the project solution is presented as an extension of a well-running ecosystem OpenDESC.com. The background to the emergence of OpenDESC.com, the customer base and the corresponding offering are described. The method how to systematically plan changes, improvements and extensions by enterprise architecture integration is discussed. The extended offering for new customer categories based on the out-come of the project is classified and described. The method for assessment of this ecosystem is proposed. Finally, conclusions and outlook summarize the insights and give future directions.

In Chap. 11, the outcome of project is given and the future perspectives drawn. The concept of Digital Twin is almost twenty years old, posing an enhancement of well-known concepts like CIM and PLM, with expected benefits through ubiquitous simulation, real-time analysis, and synchronous processing. The theoretical framework and practical implementations of Digital Twin still do not follow this vision rather discover some specific characteristics of the Digital Twin. Although many successful implementations exist, Digital Twin is still offered and sold as a project for a company's specific purpose. While sufficient implementation details are not publicly available, it is difficult to assess effectiveness of different solutions and make comparisons in a structured manner. While a standard taxonomy for Digital Twin does not exist, a recent taxonomy from literature is used to explain the state of development of Digital Twin. At first, the way for the further development of the presented research project solution is discussed. At second, current trends and challenges related to Digital Twin and its research domains are explained in a comprehensive overview. The assessment is based dimensions and characteristics which serve as an initial part of a future universal reference framework. The Digital Twin applications in different domains are classified and the possible Digital Twin evolution is discussed. Novel research trends and challenges are identified, advancing the theory and practice of Digital Twins. The results show an evolution of Digital Twin's role from an enabler of cyber-physical systems to a product lifecycle data integration and processing platform.

1.6 Contributors of the Book

Apart of their own contribution, the editors have invited the contributors based on their contribution to the research project. The singular chapters represent the authors' contribution to the dedicated work package. The editors are grateful to all contributors for their excellent work.

References

1. Grieves M, Vickers J (2017) Digital twin: mitigating unpredictable, undesirable emergent behavior in complex systems. In: Kahlen FJ, Flumerfelt S, Alves A (eds) Transdisciplinary perspectives on complex systems: new findings and approaches. Springer International, Cham, pp 85–113
2. Sobolewski M (2017) Amorphous transdisciplinary service systems. Int J Agile Syst Manage 10(2):93–115
3. Stjepandić J, Wognum N, Verhagen WJC (2015) Concurrent engineering in the 21st century. Springer International Publishing, Cham

4. Stark J (2019) Product lifecycle management: paradigm for 21st century product realisation, 4th edn, vol 1. Springer International Publishing, Cham. https://doi.org/10.1007/978-3-030-28864-8.
5. Stjepandić J, Wognum N, Verhagen WJC (2019) Systems engineering in research and industrial practice. Springer International Publishing, Cham
6. Sharma A, Kosasih E, Zhang J, Alexandra A, Calinescu A (2020) Digital twins: state of the art theory and practice, challenges, and open research questions. arXiv:2011.02833v3.
7. Wagner L (2020) Kein digital twin ohne digitale Durchgängigkeit. Zeitschrift für den wirtschaftlichen Fabrikbetrieb, Band 115, April 2020, pp 43–45. https://doi.org/10.3139/104.112315.
8. Zhang C, Chen Y (2020) A review of research relevant to the emerging industry trends: industry 4.0, IoT, blockchain, and business analytics. J Indus Integr Manage 5(1):165–180
9. De Modesti PH, Carvalhar Fernandes E, Borsato M (2020) Production planning and scheduling using machine learning and data science processes. Adv Transdisciplinary Eng 13:155–166
10. Wang Y, Wang S, Yang B, Zhu L, Liu F (2020) Big data driven hierarchical digital twin predictive remanufacturing paradigm: architecture, control mechanism, application scenario and benefits. J Cleaner Prod 248:119299.
11. Lo CK, Chen CH, Zhong RY (2021) A review of digital twin in product design and development. Adv Eng Inform 48:101297
12. Sieg O, Völl C (2020) Digital twin and Product Lifecycle Management (PLM) as part of the corporate strategy [Digital Twin und PLM als Teil der Unternehmensstrategie]. ZWF Zeitschrift fuer Wirtschaftlichen Fabrikbetrieb 115:38
13. Lu Y, Liu C, Wang KIK, Huang H, Xu X (2020) Digital twin-driven smart manufacturing connotation, reference model, applications and research issues. Robot Comput Integr Manuf 61:101837
14. Cheng J, Zhang H, Tao F, Juang CF (2020) DT-II-Digital twin enhanced Industrial Internet reference framework towards smart manufacturing. Robot Comput Integr Manuf 62:101881
15. N.N., DigiTwins, DIGITAL TWINS FOR BETTER HEALTH (2021). https://www.digitwins.org/. Accessed, 30 Jan 2021
16. Takahashi K (2020) Social issues with digital twin computing, NTT Technical Review. https://www.ntt-review.jp/archive/ntttechnical.php?contents=ntr202009fa5.html. Accessed 30 Jan 2021
17. Pobuda P (2020) The digital twin of the economy: proposed tool for policy design and evaluation. Real-World Econ Rev 94:140–148. http://www.paecon.net/PAEReview/issue94/Pobuda94.pdf. Accessed 30 Jan 2021
18. Wognum N, Bil C, Elgh F, Peruzzini M, Stjepandić J, Verhagen WJC (2019) Transdisciplinary systems engineering: implications, challenges and research agenda. Int J Agile Syst Manage 12(1):58–89
19. Wognum N, Bil C, Elgh F, Peruzzini M, Stjepandić J, Verhagen W (2018) Transdisciplinary engineering research challenges. Adv Transdisciplinary Eng 7:753–762
20. Denkena B, Stobrawa S, Sommer M, Stjepandic J, Von Soden M (2020) Production planning using the digital twin | [Effiziente Erstellung eines digitalen Zwillings zur Verwendung im Produktionsmanagement; Produktionsplanung mit dem digitalen Zwilling]. WT Werkstattstechnik 110(10):661–665
21. Sommer M, Stjepandić J, Stobrawa S, von Soden M (2019) Automatic generation of digital twin based on scanning and object recognition. Adv Transdisciplinary Eng 10:645–654. https://doi.org/10.3233/ATDE190174

Chapter 2
Requirements for the Optimization of Processes Using a Digital Twin of Production Systems

Sebastian Stobrawa, Berend Denkena, Marc-André Dittrich, and Moritz von Soden

2.1 Introduction

Production planning and control in industrial companies as part of production management includes many tasks that are performed to optimize the production system. The task of production planning and control is to achieve various logistical and economic objectives of industrial production operations. Such objectives can be quantified by target figures like delivery reliability, efficiency, stock levels, and so on [1]. To achieve these targets, production planning and control encompasses the entire production process, from order acquisition to production, warehouse and assembly to dispatch [2]. The Digital Factory supports companies in production planning and control throughout the production process in order to achieve the above-mentioned targets. It is defined as an overall term for a comprehensive network of digital models, methods and tools that are integrated by a continuous data management [3]. A central tool of the Digital Factory is simulation, which contributes to the planning and improvement of all structures, processes, and resources in real production [4]. In the field of production and logistics, simulation has been scientifically investigated and established for a long time [5–7]. However, simulation models are usually designed

S. Stobrawa (✉) · B. Denkena · M.-A. Dittrich
Institut Für Fertigungstechnik Und Werkzeugmaschinen, Leibniz Universität Hannover, An der Universität 2, 30823 Garbsen, Germany
e-mail: stobrawa@ifw.uni-hannover.de

B. Denkena
e-mail: denkena@ifw.uni-hannover.de

M.-A. Dittrich
e-mail: dittrich@ifw.uni-hannover.de

M. von Soden
Bornemann Gewindetechnik GmbH & Co. KG, Klus 3, 31073 Delligsen, Germany
e-mail: m.vonsoden@bornemann.de

J. Stjepandić et al. (eds.), *DigiTwin: An Approach for Production Process Optimization in a Built Environment*, Springer Series in Advanced Manufacturing,
https://doi.org/10.1007/978-3-030-77539-1_2

for a specific application. This could be, for example, the analysis of stocks. If a further issue is to be solved with this simulation model, this is usually only feasible with accordingly adapted simulation parameters. Therefore, it is useful to create a framework for simulation models with which these models can be used for several use cases and can be extended as needed. By using such a framework and by enabling a two-way exchange of information and control between the real and virtual world, these simulation models become a Digital Twin of production systems, which is generally defined as the digital representation of an object or system which includes characteristics, properties, conditions, and behaviours by means of models, information and data [8]. As a result, the Digital Twin provides an image of the production process, which enables the optimization of company production and the simulation of various processes in advance with the help of the digital tool. Overall, the Digital Twin is a considerable asset for companies which fosters objective-orientated production [4]. To achieve this in practice, however, the above-mentioned use cases' specific simulation models have to be adapted to a framework by selecting and specifying the respective parameters, so that an extension to a Digital Twin can be achieved.

2.2 Investigated Use Cases

In this section, the use cases that are to be investigated further are presented. From this selection, requirements and basic conditions for the further procedure arise. The application of simulation in the Digital Factory or as the basis of the Digital Twin is generally possible in a comprehensive manner so that a selection must be made from the multitude of conceivable use cases. In particular, this excludes applications which, despite promising benefits, concern other areas as well, such as energy management, personnel management or similar [9]. Of course, the procedure presented in this book can also be utilized to address these topics. For this purpose, the detailed description presented here provides a suitable pattern to be transferred to other application cases.

Factory planning: This use case consists of the reorganization of structures based on new concepts like push or pull, layout planning, or material flow analyses. Accordingly, this task mainly addresses the spatial arrangement of the factory building. One important parameter to perform the task is the location data of the relevant objects of the production system. These objects will be named and explained in the following. In addition to the spatial location, information about the production process is also relevant; with this, the position of the object can be optimized in terms of the production process.

Investment planning: An investment is a long-term capital expenditure in assets. For the application case described here, planning investments means that capital flows into capacity expansions with the aim of achieving a long-term return on these investments. This requires an economic consideration of the investment so that the costs do not exceed the expected benefit. Investment goods that are considered at this point are, in particular, new machines, equipment, or technology. For the application case, this means that the investment goods are inserted into the considered production system and the effect on the production system is examined. In addition to the spatial information, since the investment goods must also be physically inserted into the system, the production-related features must also be determined and included in the analysis.

Capacity planning: The planning of capacities for a production system determines the necessary capacity of the potential factors for the execution of the production schedule in a reference period. Potential factors in this context are machines, equipment and workers. For the application case, capacity requirements of the production system are compared with existing available capacities. The aim is to resolve this trade-off at the lowest possible cost. To fulfil the use case, information about the considered potential factors and their effects on the production system is required.

Bottleneck analysis: Production systems consist of multi-stage production steps which are interlinked and performed on different stations. The linking of these process steps causes bottlenecks in the production process, for example, due to different process step durations or malfunctions. Correspondingly, bottleneck stations arise which, subsequently, limit the entire production system. In the use case *bottleneck analysis*, these bottleneck stations are identified so that measures can be taken to relieve them and thus increase the performance of the entire system. To perform the use case, information on the production process and the interlinking of the stations is required.

Inventory management: In the use case *bottleneck analysis*, it has already been explained that production systems are characterized by interlinked production steps, but that these cannot usually be carried out synchronously due to different process or setup times. To enable synchronicity, stocks are applied. Intermediate products are temporarily stored here so that the single process steps can be loaded as high as possible. However, stocks are a high cost factor, so they should be avoided if possible. For this reason, a trade-off is reached in which there is an optimal point that, on the one hand, ensures the reliability of the production process and the synchronicity of the process steps and, on the other hand, minimizes inventory costs. This optimal point is determined with inventory planning. For this planning, information on the storage locations and the production process is required.

In-house material transport: Material transport within a production system is necessary to move products or load carriers between fixed stations and storage facilities. These processes are usually time-consuming and require a lot of space in the production system due to the routes needed. For this reason, there is a great potential for optimization in the planning of transportation. This task is performed by the internal material transport. For this application, information is required regarding the moving objects in addition to the spatial arrangement of the objects in the given production system. Information about the production process is also needed.

Figure 2.1 summarizes the discussed use cases. In the framework presented here, the production system to which the use cases are applied is represented by the parameters introduced in more detail in the following section. These parameters determine

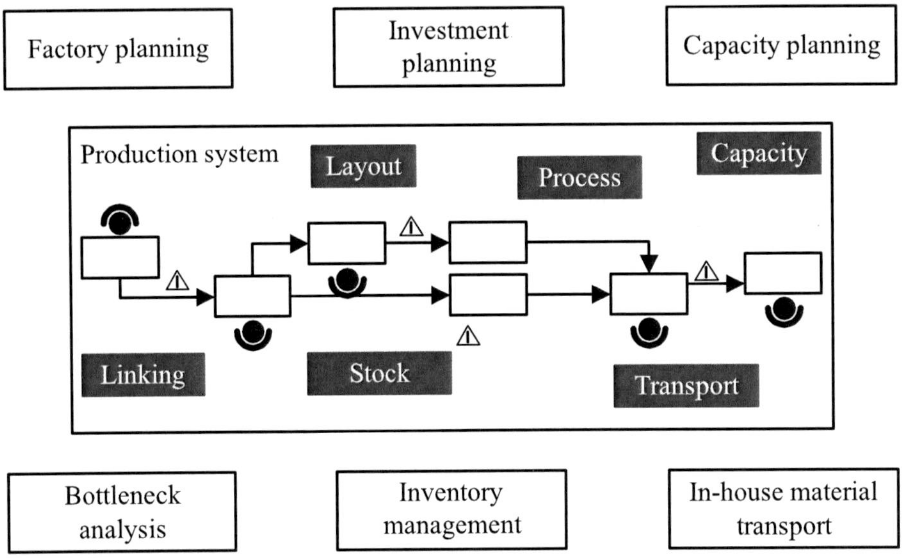

Fig. 2.1 Use cases and related objects

the information defining layout, process, capacity, linking, stock, and transport. This information is necessary to process the use cases as further explained in the following. For this purpose, the objects are examined in the next section.

2.3 Parameters for the Framework of the Digital Twin

In general, the spatial position and orientation of the entities of a production system are relevant for the Digital Twin. On the one hand, this allows correct mapping of movements within the production system and calculation of transport times for individual objects. On the other hand, the correct arrangement of the elements makes the representation directly understandable and interpretable. This is important for animations, for example, but also for spatial planning within factory planning.

The most important elements of a production system, which are also mapped in the framework for Digital Twin, are:

Stations: This term includes all objects in which value creation in a narrow sense occurs, such as machine tools, assembly stations, automated systems or robots, and rework stations. Here, the single production steps are executed on the respective product. Therefore, processing time is spent at these objects. Depending on the given production system, set-up times are also mapped here. In addition, malfunctions can cause delays in the production process, for which reason these malfunctions are also mapped. In general, it is necessary for each application to include stations in simulations because the production process to be examined cannot be modelled without them.

Products: This object type includes the raw materials, additives, intermediate and finished products which move between the stations, and thus the moving elements of the production system. However, the products can also be represented in simplified form as production orders. The only important thing at this point is that the elements each trigger a specific process at the stations. The products are also necessary for each mentioned application of the simulation, as they determine the production process according to their work plan.

Load carriers: Similar to products, load carriers are also moving elements in the production system. The special feature which distinguishes them from the products is that these products are transported in the load carriers. These objects include, for example, boxes, pallets or even vehicles. The modelling of load carriers is useful in factory planning if they have a relevant influence on size such that the storage of the load carriers must be taken into account. They can also be included in bottleneck analyses, inventory management, and material transport if the level of detail justifies the modelling effort. It should be emphasized that, in many cases, load carriers can be represented by products in a simplified way, so that the listing of this element is only given for the sake of completion.

Warehouse: This element refers to all elements of a production system which store products for longer periods of time. This includes incoming, outgoing, and intermediate storage. Warehouses are not explicitly integrated into the process flow. For this reason, the modelling of warehouses is only necessary if their size is relevant for factory planning, if the stock contained in them is to be considered in inventory planning, or if the material transport should include the route to and from the warehouse. The level of detail varies depending on the specific application in which the warehouse can be modelled, simply as a fixed location or single spaces in detail.

Buffer: The elements which are assigned to the object *buffer* store products, as mentioned in the object type warehouses. However, in contrast to warehouses, buffers store products temporarily and close to the stations. Therefore, buffers are also integrated into the process flow. The types of buffers include all storage locations where products are placed before or after a station. For example, these can be floor areas or supply shelves. Buffers are relevant for factory planning because they occupy space in the production hall, for bottleneck analysis, as stations can be synchronized with them, for inventory planning, because they contain work-in-progress stock, and for material transport, because products are stored or picked up at these locations.

Workplaces: These elements are stations where manual tasks are performed. A mapped object is always required where the operations are executed, for example, stations (for processing or set-up tasks) or buffers, warehouses, and conveyors (for transport tasks). Workstations are necessary to be able to integrate a working person into the Digital Twin, which, in turn, is necessary for the above-mentioned applications should the level of detail be requested. It should therefore be emphasized that workplaces can be modelled in a simplified way by using the information in the stations, for example, by adjusting the processing times accordingly. However, simulations of labour-intensive production systems typically have to be modelled with human resources.

Employees: In addition to the workstations, the working persons themselves also play a role regarding manual labour. Accordingly, they constitute objects in the Digital Twin. The degree of detail depends on the intended use case, so that properties of the employees, such as work efficiency and speed, can be integrated as needed. The mapping of employees is required for all applications, as described in the section on workplaces, if a labour-intensive production system is given.

Conveyors: The objects refer to all (partially) automated installations for transporting products or load carriers. This includes conveyor belts, roller belts, cranes and so on. Material transport within the production system is performed with regard to specific transport times. The objects of the conveyors are necessary for all applications in which the transport within the production system represents a critical time expenditure. Accordingly, there is the possibility here that the transports are modelled in a simplified way if the transport within the production system plays a subordinate role.

Transport routes and paths: Within a production system, defined paths are used to move material, people, or other items. These routes are also objects which are modelled in the Digital Twin in order to be able to implement the movements in the simulation—for example, for the determination of transport times. This is relevant for all applications because the production process is determined by it, but especially for factory planning, bottleneck analysis and material transport.

The shown objects are now further defined by assigning properties. These properties include: position data, output behaviour, time portion, setup information, malfunction information, working hours, and type-specific properties. At this point, it should be emphasized that this is an initial generation, so that e. g. the position data of products is therefore not applicable. Table 2.1 presents an overview of the relevant properties of the respective objects. A detailed table of all used parameters can be found in the appendix. In order to make the procedure comprehensible at this point, the object "station" is explained in detail in the following section.

2.4 Parameter Properties for Implementation in the Digital Twin

It has been shown that the execution of different use cases requires appropriate parameters, as detailed in Sects. 2.2, 2.3. To further define the use cases, these parameters are described in more detail in this section. For the generation of a Digital Twin, these parameters must be available in a defined way. For this purpose, the parameters shown in Table 2.1 are assigned to each object. Further, each parameter is assigned

Table 2.1 Properties of the objects

Objects	Position data	Output behaviour	Times portion	Malfunction	Working hours	Object-specific properties
Stations	✔	✔	✔	✔	✔	Setup information
Products						Dimensions, batch size
Load carriers						Capacity, dimensions
Warehouse	✔			✔		Capacity
Buffer	✔	✔	✔	✔		Capacity
Workplaces	✔					Assigned object, tasks
Employees				✔	✔	Efficiency, speed, tasks
Conveyors	✔	✔	✔	✔		Capacity, stowage, backwards
Routes	✔			✔		Width, capacity

based on the way the acquisition is performed, which data type is used, whether the parameter is mandatory or optional, for which use case the parameter is needed, and whether the parameters must be recorded once or continuously. The procedure is explained in more detail using the example of a station.

Initially, each object must have a label, which corresponds to the real name, and another unique identification (Table 2.2). This information is only used for the identification in the simulation model and the software implementation. Beyond that, Table 2.1 indicates that position data, output behaviour, time portion, malfunctions, working hours, and setup information are required to define the stations for the simulation.

The position data consists of the spatial information in X, Y and Z position (Table 2.3). This is used to locate the station spatially in the factory hall so that the layout is correctly included in the model. It is possible to specify either the centre of the station in the three spatial coordinates or a bounding box. Furthermore, the

Table 2.2 Basic information of stations

Basic information		
Parameter	Name	ID
Acquisition	Manually	Automatically
Data type	String	String
Mandatory or optional	Mandatory	Mandatory
Assigned use cases	All	All
Recording	Once	Once

Table 2.3 Position data of stations

Position data			
Parameter	X-coordinate	Y-coordinate	Z-coordinate
Acquisition		Scan	
Data type		Real	
Mandatory or optional	Mandatory		Optional
Assigned use cases		All	
Recording		Once	

position data is either acquired by a scan or entered manually. The input is done via a numeric data type, for example real. Since the position data determines the shape of the simulation model and a Digital Twin can only be interpreted in this way, the position data is mandatory and relevant for all use cases. The Z-position, i.e. the position in height, is not relevant in many cases because all objects are on one level. Therefore, the Z-coordinate is optional. The recording is performed once and only needs to be updated if there are any changes in the shop floor.

The stations are connected in the material flow, which means that the products move from one station to the next according to a certain logic, usually the work plan. These connections can either result directly from the work plan or be predetermined in the case of rigidly linked production systems. In the second case, an output behaviour is then stored in the station properties (Table 2.4). For example, these can be simple rules such as distributing the products evenly to all downstream stations or a certain percentage distribution. The output behaviour is entered manually and described with the data type string. This is optional data but, if applied, it is relevant for all applications, as it determines the material flow. The recording is necessary only once.

The time portion indicate the time sequence at the stations (Table 2.5). The following portion are distinguished:

- Processing time: The time in which the actual process is performed. This time portion is mandatory for the Digital Twin, as it defines the process and is therefore essential for the material flow.

Table 2.4 Output behaviour

Output behaviour	
Parameter	Output behaviour
Acquisition	Manually
Data type	String
Mandatory or optional	Optional
Assigned use cases	All
Recording	Once

Table 2.5 Time portion

Time portion		
Parameter	Times share	Distribution
Acquisition	Manually/database	Manually/database
Data type	Time	String
Mandatory or optional	Mandatory/optional	Optional
Assigned use cases	All	All
Recording	Continuously	Continuously

- Recovery time: Time required to prepare the single process, e.g. loading a part into the machine (this definition is taken from the simulation software). This is an optional specification.
- Cycle time: This optional time portion defines the cycle of a station. With this, processes controlled by the stroke can be modelled.
- Set up time: This is used to define the duration of a setup process. This optional information is specific to stations and is not included in other objects.

For each time portion described by the datatype time, a statistical distribution (datatype string) is optionally recorded to be able to map fluctuations in the process. The recording of the data is usually done manually, but it can be automated by accessing data stored in databases. This can be enterprise resource planning (ERP) and manufacturing execution systems (MES). Also, this can be the scan of the objects compared with a corresponding reference database in which station data is stored. For this purpose, the currently existing work from Denkena et al. [10] can be considered. Time portion are used in all use cases because they determinate the process modelled. The recording is done continuously because current and correct time portion determine the validity of the Digital Twin and large fluctuations or changes can occur.

In many cases, the process flow in a production system can only be modelled correctly if malfunctions of the stations are also mapped. This allows valid analysis results, especially in bottleneck analysis and inventory management. The parameters, however, are generally optional, since analyses can be performed without them, e.g. in production systems where malfunctions have no major influence (Table 2.6). The malfunctions for stations are integrated by two parameters: mean time to repair (MTTR) and availability. The MTTR indicates how long on average a disturbance lasts until it is rectified, the data type time is used accordingly. The availability indicates the percentage ratio of the disturbance-free time to the total time and is thus indicated with the numerical data type real. Both parameters must be entered manually or obtained from a database, as with the time portion. The recording of the parameters should always be renewed in order to register changes and keep the Digital Twin valid [11].

Table 2.6 Malfunction

Malfunction		
Parameter	MTTR	Availability
Acquisition	Manually/database	Manually/database
Data type	Time	Real
Mandatory or optional	Optional	Optional
Assigned use cases	Bottleneck, inventory	Bottleneck, inventory
Recording	Continuously	Continuously

Table 2.7 Working hours

Working hours	
Parameter	Working hours
Acquisition	Manually
Data type	Table
Mandatory or optional	Optional
Assigned use cases	Capacity planning
Recording	Once

The last property of stations considered is the working time (Table 2.7). This indicates the times during the working day when the stations are operating. Accordingly, breaks and rest periods are considered. The working hours are listed in a table which must be created manually. Working hours are generally optional. They are not explicitly necessary in cases without an integrated production schedule, but only a general process sequence is examined. However, for capacity planning, the integration of working hours is recommended. Since the working hours do not usually change, this must be included only once.

With the mentioned properties, stations are described with sufficient accuracy for a Digital Twin of a production system. For all other objects, corresponding investigations have been performed. The results can be found in the appendix. At this point, it is emphasized that the implementation of these parameters in a Digital Twin is aimed at the use in the simulation software Plant Simulation by Siemens. If another simulation software is used, it should be checked how efficiently the single parameters can be mapped. However, a rough examination of comparable simulation programs led to the conclusion that the parameter selection is generally transferable.

2.5 Summary and Outlook

In this chapter, it was described how a Digital Twin of production systems can be parameterized by considering relevant use cases. This procedure was necessary to restrict the investigation scope because many other parameters can be included for other use cases. For example, energy efficiency analyses would require additional objects and parameters. Furthermore, it was explained how the individual resulting objects are structured into parameters. The efficient determination of these parameters via scan, object recognition and forms are outlined in the following chapters of this book. In the next chapters, the parameters have been further examined so that the data type, the type of capture, and the repeat for each parameter have been determined. The explanations thus form the basis for an efficient generation of a Digital Twin.

Appendix

These abbreviations apply to the following tables:

- Acquisition group:
  1—Automated recording via a scan or a database extract
  2—Semi-automated acquisition via data post-processing, e.g. object recognition
  3—Manual recording using expert queries or forms.
- Data type:
  Boolean—True or false
  Integer—Numeric, positive number
  Object—Reference to another object
  Real—Numeric; negative and positive real numbers
  String—Any characters
  Table—Two-dimensional specification of strings
  Time—Time specification, numeric in seconds (optional).
- Use case:
  1—Factory Planning
  2—Investment planning
  3—Capacity planning
  4—Bottleneck analysis
  5—Inventory management
  5—Inventory management
  6—In-house material transport

Parameter	Description	Acquisition group	Data type	Mandatory data	Use case	Recording
Stations						
Name	Name of the station	1/3	String	Yes	All	Once
ID	Unique naming	1/3	String	Yes	All	Once
X-coordinate	Length of the product in X-dimension	1	Real	Yes	All	Once

(continued)

(continued)

Parameter	Description	Acquisition group	Data type	Mandatory data	Use case	Recording
Y-coordinate	Length of the product in Y-dimension	1	Real	Yes	All	Once
Z-coordinate	Length of the product in Z-dimension	1	Real	Yes	All	Once
Output behaviour	According to which principle is the product sent to the next station	3	String	No	All	Once
Processing time	Default processing time	2/3	Time	Yes	All	Continuously
Processing time distribution	Statistical distribution of the processing time	2/3	String	No	–	Once
Setup time	Default setup up time	2/3	Time	No	All	Continuously
Setup time distribution	Statistical distribution of the setup time	2/3	String	No	–	Once
Cycle time	Default recovery time	2/3	Time	No	–	Continuously
Cycle time distribution	Statistical distribution of the recovery time	2/3	String	No	–	Once
Automatic setup	Specifies whether the setup should be performed automatically	3	Boolean	No	–	Once
Set up by number of parts	Setup after a certain number of parts	3	Integer	No	–	Once
Set up by attribute	Setup according to a specific product attribute (e.g. name)	3	String	No	–	Once

(continued)

(continued)

Parameter	Description	Acquisition group	Data type	Mandatory data	Use case	Recording
Mean time to repair	Average malfunction duration	2/3	Time	No	–	Continuously
Availability	Average percentage of the functionality of the station	2/3	Real	No	–	Continuously
Shift calendar	Times at which production takes place (minus breaks and rest periods)	3	String	No	All	Once
Products						
Name	Name of the worker	1/3	String	Yes	All	Once
ID	Unique naming	1/3	String	No	–	Once
Width	Width of the object	1/3	Real	No	5, 6	Once
Length	Length of the object	1/3	Real	No	5, 6	Once
Height	Height of the object	1/3	Real	No	5, 6	Once
Batch size	Number of Products, that can be summarized as a unit	1/3	Real	No	all	Once
Load carriers						
Name	Name of the load carrier	1/3	String	Yes	All	Once
ID	Unique naming	1/3	String	No	–	Once
Capacity	Number of units, that can be stored in one load carrier	1/3	Integer	No	5, 6	Once
Width	Width of the Load carrier	1/3	Real	No	5, 6	Once
Length	Length of the Load carrier	1/3	Real	No	5, 6	Once

(continued)

(continued)

Parameter	Description	Acquisition group	Data type	Mandatory data	Use case	Recording
Height	Height of the Load carrier	1/3	Real	No	5, 6	Once
Warehouse						
Name	Name of the warehouse	1/3	String	Yes	All	Once
ID	Unique naming	1/3	String	No	–	Once
X-coordinate	X-position of the object	1	Real	No	1, 6	Once
Y-coordinate	Y-position of the object	1	Real	No	1, 6	Once
Z-coordinate	Z-position of the object	1	Real	No	1, 6	Once
Capacity	Number of units, that can be stored in the warehouse	2/3	Integer	No	3, 5	Once
Mean time to repair	Average malfunction duration	2/3	Real/time	No	3, 4, 5, 6,	Continuously
Availability	Information about the availability of the object	2/3	Real	No	3, 4, 5, 6,	Continuously
Buffer						
Name	Name of the buffer	1/3	String	Yes	All	Once
ID	Unique naming	3	String	No	–	Once
X-coordinate	X-position of the object	1	Real	No	1, 6	Once
Y-coordinate	Y-position of the object	1	Real	No	1, 6	Once
Z-coordinate	Z- position of the object	1	Real	No	1, 6	Once
Output behaviour	According to which principle is the product sent to the next station	3	String	No	All	Once
Capacity	Capacity of the buffer	2/3	Integer	No	3, 5	Once

(continued)

(continued)

Parameter	Description	Acquisition group	Data type	Mandatory data	Use case	Recording
Mean time to repair	Average malfunction duration	2/3	Real/time	No	3, 4	Continuously
Availability	Information about the availability of the object	2/3	Real	No	3, 4	Continuously
Workplaces						
Name	Name of the workplace	1/3	String	Yes	All	Once
ID	Unique naming	3	String	No	–	Once
X-coordinate	X-position of the object	1	Real	No	1, 6	Once
Y-coordinate	Y-position of the object	1	Real	No	1, 6	Once
Z-coordinate	Z-position of the object	1	Real	No	1, 6	Once
Tasks	Used to define various work tasks	3	String	No	–	Continuously
Workstation	Assigned single station	1/3	Object	No	–	Once
Employees						
Name	Name of the Employee	1/3	String	Yes	All	Once
ID	Unique naming	1/3	String	No	–	Once
Efficiency	Expresses how efficiently the worker works	2/3	Real	No	3, 4	Continuously
Speed	Speed at which the worker walks on paths	2/3	Real	No	3, 4, 6	Once
Shift calendar	Times at which production takes place (minus breaks and rest periods)	3	String	No	All	Once
Mean time to repair	Average malfunction duration	2/3	Real/time	No	3, 4, 6	Continuously

(continued)

(continued)

Parameter	Description	Acquisition group	Data type	Mandatory data	Use case	Recording
Availability	Information about the availability of the object	2/3	Real	No	3, 4, 6	Continuously
Conveyors						
Length	Length of the object	1	Real	No	1, 6	Once
Output behaviour	According to which principle is the product sent to the next station	3	String	No	All	Once
Mean time to repair	Average malfunction duration	2/3	Real/time	No	3, 4	Continuously
Availability	Information about the availability of the object	2/3	Real	No	3, 4	Continuously
Service	Can be used to define various work tasks (production, transport, set-up)	3	String	No	–	Continuously
Recovery time	Time that elapses before the transport starts after the product has reached the conveyor	1/3	Real/time	No	–	Continuously
Cycle time	Time specification, which indicates at what distance products can be transported	1/3	Real/time	No	–	Continuously
Automatically stop	Conveyor stops if there is no object on it	3	Boolean	No	4, 6	Once

(continued)

(continued)

Parameter	Description	Acquisition group	Data type	Mandatory data	Use case	Recording
Backwards	With this setting the conveyor can also run backwards	1/3	Boolean	No	4, 6	Once
Capacity	Number of simultaneously transportable units	1/2/3	Integer	No	3, 5, 6	Once
Distance	Difference between two products	3	Integer/real	No	4, 6	Once
Time	Duration of transport on the conveyor	1/3	Time	Yes	3, 4, 5, 6	Continuously
Speed	Transport speed of the conveyor	2/3	Real	Yes	3, 4, 5, 6	Continuously
Acceleration	Acceleration of the conveyor	2/3	Real	No	6	Continuously
Delay	Stopping behaviour of the conveyor	2/3	Real	No	6	Continuously
Assigned workstations	Assigns a workstation to a machine	1/3	String/object/	No	–	Once
Routes						
Name	Name of the route	1/3	String	Yes	All	Once
ID	Unique naming	3	String	No	–	Once
Width	Width of the transport route	1	Real	No	1, 6	Once
Capacity	Number of units that can be transported simultaneously	1/3	Integer	No	3, 5, 6	Once
Mean time to repair	Average malfunction duration	2/3	Real/time	No	1, 6	Continuously
Availability	Information about the availability of the object	2/3	Real	No	1, 6	Continuously

References

1. Nyhuis P, Wiendahl HP (2009) Fundamentals of production logistics. Springer, Berlin Heidelberg
2. Hackstein R (1984) Produktionsplanung und -steuerung (PPS) - Ein Handbuch für die Betriebspraxis. VDI Verlag, Düsseldorf
3. VDI 4499 (2008) Digitale Fabrik. Beuth Verlag, Berlin
4. Wenzel S, Peter T (2017) Prozess-Simulation in kleinen und mittleren Unternehmen mittels des Bausteinkastensystems KMUSimMetal. In: Wenzel S, Peter T (eds) Simulation in Produktion und Logistik. University Press, Kassel, pp 433–442
5. Spieckermann S (2005) Diskrete, ereignisorientierte Simulation in Produktion und Logistik - Herausforderungen und Trends. Simulation und Visualisierung 2005 - SimVis, pp 3–14
6. Rabe M, Spieckermann S, Wenzel S (2008) Verifikation und Validierung für die Simulation in Produktion und Logistik: Vorgehensmodelle und Techniken. Springer, Berlin
7. Wenzel S, Weis M, Collisi-Böhmer S, Pitsch H, Rose O (2008) Qualitätskriterien für die Simulation in Produktion und Logistik. Springer, Berlin
8. Stark R, Damerau T (2019) Digital twin. In: Chatti S, Laperrière L, Reinhart G, Tolio T (eds) The international academy for production engineering. CIRP encyclopedia of production engineering. Springer, Berlin Heidelberg
9. Denkena B, Dittrich MA, Stobrawa S, Stjepandic J (2019) Automated generation of a digital twin using scan and object detection for data acquisition. 18. ASIM Fachtagung: Simulation in Produktion und Logistik, 18.-20. September 2019, Chemnitz, Wissenschaftliche Scripten, Auerbach, pp 49–58
10. Denkena B, Doreth K, Köller M, Wilmsmeier S, Winter F (2020) Simulationsbasierte kombinierte Instandhaltungs- und Produktionsplanung. In: Mayer G, Pöge C, Spieckermann S, Wenzel S (eds) Ablaufsimulation in der Automobilindustrie. Springer Vieweg, pp 261–273
11. Denkena B, Dittrich MA, Keunecke L, Wilmsmeier S (2020) Continuous modelling of machine tool failure durations for improved production scheduling. Prod Eng Res Develop (WGP) 14(2):207–215

Chapter 3
Digital Twin: A Conceptual View

Josip Stjepandić, Markus Sommer, and Sebastian Stobrawa

3.1 Introduction

In the past decade, the term Digital Twin was coined to describe the entire virtual representation of a physical system, process or service. In the beginning, Digital Twins were merely descriptive, but as computational, information and communication technologies evolved, a high-performance bidirectional coupling between the digital and the physical system was established [1]. Digital Twin was meant to improve competitiveness of the industry, especially in manufacturing, to meet challenges caused by volatile demand, even lower batch size and high cost pressure [2].

After few years of tremendous development, academia has recently announced the following definition of Digital Twin [3]: "A digital twin is a digital representation of an active unique product (real device, object, machine, service or intangible asset) or unique product service system (a system consisting of a product and a related service) that comprises its selected characteristics, properties, conditions and behaviours by means of models, information and data within a single or even across multiple lifecycle phases".

J. Stjepandić (✉)
PROSTEP AG, Dolivostraße 11, 64293 Darmstadt, Germany
e-mail: josip.stjepandic@prostep.com

M. Sommer
isb innovative software businesses GmbH, Otto-Lilienthal-Strasse 2, 88046 Friedrichshafen, Germany
e-mail: sommer@isb-fn.de

S. Stobrawa
Institut Für Fertigungstechnik Und Werkzeugmaschinen, Leibniz Universität Hannover, An der Universität 2, 30823 Garbsen, Germany
e-mail: stobrawa@ifw.uni-hannover.de

J. Stjepandić et al. (eds.), *DigiTwin: An Approach for Production Process Optimization in a Built Environment*, Springer Series in Advanced Manufacturing,
https://doi.org/10.1007/978-3-030-77539-1_3

The Digital Twin is based on an experiment-capable digital model, thus a simulation model. This model provides a virtual representation of the real system or object. Here, only those properties of the real system are mapped in the digital model that lead to accurate virtual representation of the system behaviour [4]. The specific system behaviour can therefore be abstracted and understood. With the digital model, it is possible to conduct analyses and evaluations that were previously solved either by expert knowledge or mathematical calculations [5]. However, since the digital model can be applied to integrate complex interdependencies, statistical distributions and scenarios into the investigations, new evaluation and analysis possibilities emerge. These potentials will be further enhanced if the digital model is extended to a Digital Twin [6]. This occurs in particular when the Digital Twin interacts with the real system [7]. As a result, analyses and evaluations become more reliable, faster, and the Digital Twin provides decision-making for complex planning and control problems [8]. The benefits of Digital Twin therefore lie in an increased accuracy and fidelity as well as decreased time and costs [9]. The workload in the process of creating real-time virtual representation or in the process of realising virtual descriptions in the physical system is also decreased [1].

Several new terms were used to indicate more specific approaches of Product Lifecycle Management (PLM) [10], like Digital Thread [11], Digital Shadow [5], and Digital Twin. These three terms are partially used as synonyms and hardly to distinguish. *Digital Thread* is the overarching term which comprises a continuous connection of all digital models over the entire product lifecycle phases and all involved IT systems and databases, enabling a traceability from requirements until retirement [11]. The *Digital Shadow*, in contrast, provides a similar approach to the Digital Twin [5]. However, the Digital Shadow is considered to be a digital model that is continuously updated by a connection to the real system. A data transmission to the real counterpart is not implemented. Therefore, the Digital Shadow is not able to intervene in a controlling way.

The key criterion for differentiation between these similar terms is the level of data integration [12]. Assumed the highest level of integration, the Digital Twin is a virtual dynamic representation of a physical system, which is connected to it over the entire lifecycle for bidirectional data exchange [13, 14]. The Digital Shadow, therefore, can be understood as a lower implementation level of the Digital Twin and the Digital Thread as an approach or method to accomplish the Digital Twin.

The reason why Digital Twins are becoming more and more endorsed by industry is primarily caused by their two characteristics: their ability to integrate large amounts of static, real-time, structured and unstructured data and to combine this data with advanced data processing methods such as artificial intelligence (AI) [15], machine learning (ML) [16, 17] or high-performance computing (HPC) [18] in order to provide simulation, control [19] and self-improvement [20].

Hence, such an overwhelming term and definition needs a structure as subdimensions of the term Digital Twin. While this concept evolved from the field of PLM, the most helpful way would be a breakdown of the term according to the phases of the product lifecycle [21]. This is also in line with parts of the literature [9, 18]. Thus, the definition was split into three subtypes of Digital Twins—the Digital Master [22],

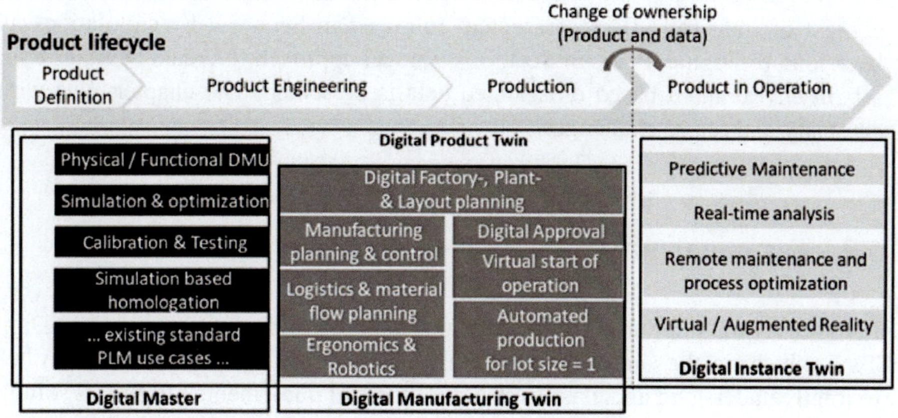

Fig. 3.1 The breakdown of the Digital Twin according to the phases of the product lifecycle [11]

the Digital Manufacturing Twin [23], and the Digital Instance Twin [24]. Digital Master addresses topics such as Functional DMU [25], function simulation [26], calibration and testing (Fig. 3.1, left) [27]. Digital Manufacturing Twin (Fig. 3.1, mid) encompasses topics such as Digital Factory [7], plant and layout planning [28], manufacturing planning [29], material flow planning [30], robotics [31] and digital approval [32]. Digital Instance Twin addresses topics such as predictive maintenance [14], real-time analysis [24] and process optimization [33] (Fig. 3.1, right).

Product Lifecycle Management combines different IT methods, architectures and systems in a uniform environment, with which products are accompanied from the first idea through development, prototyping and production to the end of use and recycling. PLM enables the continuous planning, analysis, simulation, and optimization of a product over the entire lifecycle [10]. Digital Twins leverage data from CAx systems, product lifecycle software, manufacturing systems, and sensors to create a realistic virtual model of the product, enabling prediction of performance, maintenance, and failures [34]. In such sense, PLM manages the Digital Twins. Nowadays, the products' environmental impact could be diminished by improving various product-life stages using multiple sensors due to the Industrial Internet of Things (IIOT) [35]. This would allow for more respectable methods of consumption and allow pay-to-use and product-service systems to develop further. PLM and IIoT systems are subject of more advanced integration by using Digital Thread [36].

In particular and beyond the state-of-the-art, the concept of the Digital Twins offers opportunities for business-to-business (B2B) manufacturing companies to easier co-create and maintain Product-Service-Systems [37]. This allows much closer relationship with customers base over the lifecycle of the delivered solution. For this purpose, concepts of Digital Twin and real-time simulation support each other and in the background the transformation which should start in PLM [13].

In this chapter, essential properties of Digital Twin are discussed. In Sect. 3.2, the conception of Digital Twin is briefly discussed together with its achievement as recorded in the literature. In Sect. 3.3, the main expressions of Digital Twin in context of Industry 4.0 are presented, followed in Sect. 3.4 by simulations as a backbone of Digital Twin. In Sect. 3.5, a novel approach to create Digital Twin in a built environment based on scanned data is presented. The chapter ends with a summary and future challenges.

3.2 Taxonomy of Digital Twin

In the case of a fluid, broad term without a clear scientific definition such as Digital Twin, which is at the same time used both by academia and business, a taxonomy helps to jointly understand and classify the specified and implemented technical solution. A meaningful taxonomy of Digital Twin based on 122 recent journal papers was achieved through a systematic procedure after three iterations under fulfilment of 13 ending conditions [38]. After the classification of all objects during the literature research, a concise, robust, comprehensive, extendible, and explanatory taxonomy was derived, which consists of no repetitive dimensions or characteristics. Ultimately, eight dimensions with 18 corresponding characteristics remain, which are shown in Table 3.1.

The meta-characteristics defines the taxonomy's purpose, the identification of the central, distinguishing features and properties of Digital Twins. During the development iterations, three non-mutually exclusive dimensions and five mutually exclusive dimensions were selected, as indicated in Table 3.1 [38]. Each dimension has two or three characteristics what supports clarity.

The dimension data link specifies how the communication between the Digital Twin and its physical counterpart takes place, which can either be one-directional or bi-directional. A Digital Twin can only obtain a one- or a bi-directional data link, which makes the dimension mutually exclusive [38].

The way of handling data by Digital Twin determines its overall purpose. It needs to be distinguished between three possible characteristics: processing data (1), such as monitoring, analysis, forecasting, or optimization; transfer data (2) from one point (e. g., the physical part) to another one (e.g., a data warehouse); data repository (3). A Digital Twin may have one, two or all three characteristics of this dimension at the same time. Thus, this dimension is not mutually exclusive [38].

The dimension conceptual elements describe the relationship between the Digital Twin and its physical counterpart. Whereas some authors describe a deep connection between the virtual and the physical part and some are stressing the point that the physical system is even an integral part of the whole Digital Twin, others see only a loose connection between a digital representation and its physical twin. Therefore, two characteristics for this dimension can be defined. Either a Digital Twin is directly bound to its physical part in a one-to-one ratio, or it is independent. If it is independent, a Digital Twin can be seen in combination with other physical systems or one system can possess multiple Digital Twins. This dimension is mutually exclusive [38].

Model accuracy concerns how the accuracy of the digital representation of the physical object is expressed: either by an identical accuracy or a partial accuracy.

Table 3.1 Taxonomy of Digital Twin, derived from [38]

Dimension	Characteristics			Exclusivity
Data link	One-directional		Bi-directional	Mutual
Purpose	Processing	Transfer	Repository	Not
Conceptual elements	Physically Independent		Physically Bound	Mutual
Accuracy	Identical		Partial	Mutual
Interface	M2M		HMI	Not
Synchronization	With		Without	Mutual
Data input	Raw data		Processed data	Not
Time of creation	Physical part first	Digital part first	Simultaneously	Mutual

The former describes every detail of a physical object in its digital images. It is not distinguished whether a particular detail will be relevant for the task the Digital Twin has to perform or not. Contrary to this, partial model accuracy is applied when a digital image only reflects crucial parts of the physical object. The model accuracy is mutually exclusive as well [38].

The dimension interface concerns the capability of a Digital Twin to transfer data after it processes them. After some changes to this dimension, it can be concluded that a Digital Twin could possess a machine-to-machine interface or a human–machine interface. Multiple choices are possible here: a human–machine interface via augmented reality, a machine-to-machine interface to other models or both, a human–machine interface as well as a machine-to-machine interface. Therefore, this dimension is not mutually exclusive [38].

The dimension synchronization consists of two characteristics: a working synchronization between the Digital Twin and the physical part by (real-time) data updates during its lifecycle or without a synchronization at all. The synchronization is mutually exclusive [38].

The dimension data input differentiates between raw and processed data. Digital Twins receive their data from sensors or databases. Those data might be pure, raw data gathered directly from sensors or other data collection devices. In addition, data, which are pre-processed (e. g., by analytic software) before it is transferred to the Digital Twins, might be used. Data input is not mutually exclusive.

The time of creation distinguishes between three characteristics determining the chronological order in which the respective parts of the Digital Twin come into existence. Thus, the dimension distinguishes whether the physical part or the digital part is developed first, or both parts are developed simultaneously. Most Digital Twins are designed after a physical system [38].

3.3 Conception of Digital Twin

In this section, a brief conception is sketched of the main expressions of Digital Twin in the past years. First, main subtypes of Digital Twins related to specific domains and phases of product lifecycle are described.

3.3.1 Digital Master

The Digital Master enables the enterprise to collect, maintain and provide all system information at a dedicated point in time to all actors. Downstream process can access this product information for their dedicated needs. Digital Master baselines allow traceability for all system elements. Basically, the digital master replaces a document-based approach in the development of complex products, which enables modern organization to share product data with downstream processes (Fig. 3.2). Digital

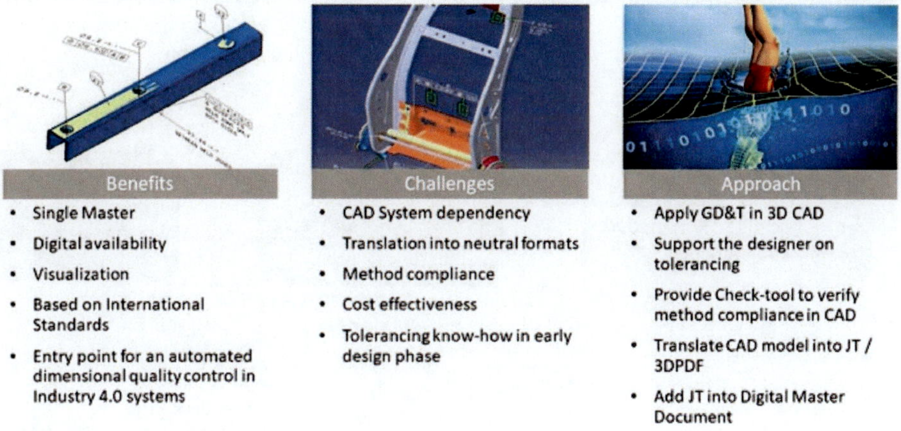

Fig. 3.2 Digital master with an application [19]

master models are a set of linked data records in self-contained document that provide a defined degree of maturity across the product lifecycle [19].

3.3.2 Digital Manufacturing Twin

Digital Manufacturing Twin is a highly accurate virtual model of the manufacturing process of a product. These models are used to simulate real-world conditions prior to generating a product or developing manufacturing operations, with the goal of optimising as much as possible in software, where multiple use cases and operation conditions are evaluated inexpensively [7].

Based on data from the digital thread, Digital Twin is constantly updated for maximum accuracy and high fidelity. By collecting information from real-life manufacturing processes, manufacturing simulations can be improved and updated, resulting in more efficient manufacturing processes (Fig. 3.3) [39]. Digital Twins are also an excellent means to capture, maintain and replicate manufacturing best practices.

3.3.3 Digital Instance Twin

The Digital Instance Twin (DIT) provide access to the product along the full post-production lifecycle (right column in Fig. 3.1). DIT is created at the end of production or delivery and may hold a copy of some of Digital Manufacturing data such as the product's configuration. If multiple instances of the same type of assets (e.g. a wind turbine) are being monitored, each of them is considered as a single Digital Instance Twin. In some sense, an aggregate of twins can be collected where similar patterns

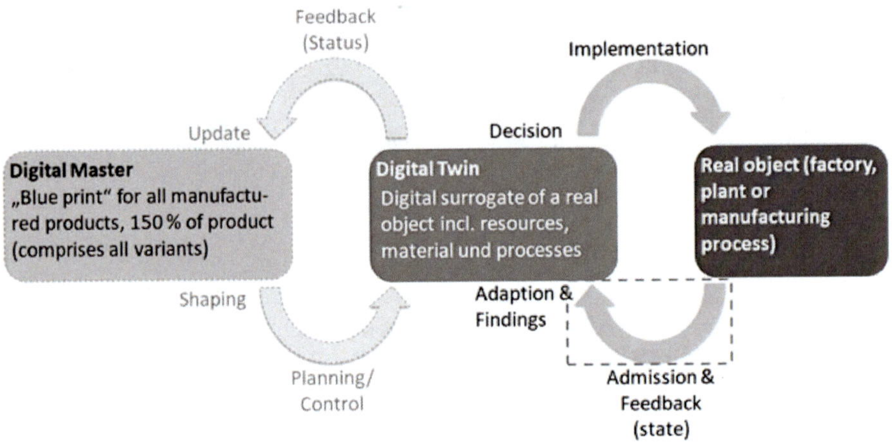

Fig. 3.3 Digital manufacturing twin [11]

or trends can be found to optimize the operation and refine models for higher fidelity in the future (Fig. 3.4).

In this way, the digital twin ecosystem is created with a central, comprehensive platform and the potential to build up several use cases and innovative business models [40].

Digital Instance Twin is used to facilitate maintenance of an operating aircraft based on its lifelong collected data. Furthermore, experts located at different sites can join via a virtual session. By inspecting a 3D model of the aircraft component, they can see synchronized information from a Digital Twin database. With Augmented Reality glasses, the Microsoft HoloLens, a Digital Twin can be experienced personally. In the inspector's view, the 3D model of the Digital Twin is directly superimposed on the

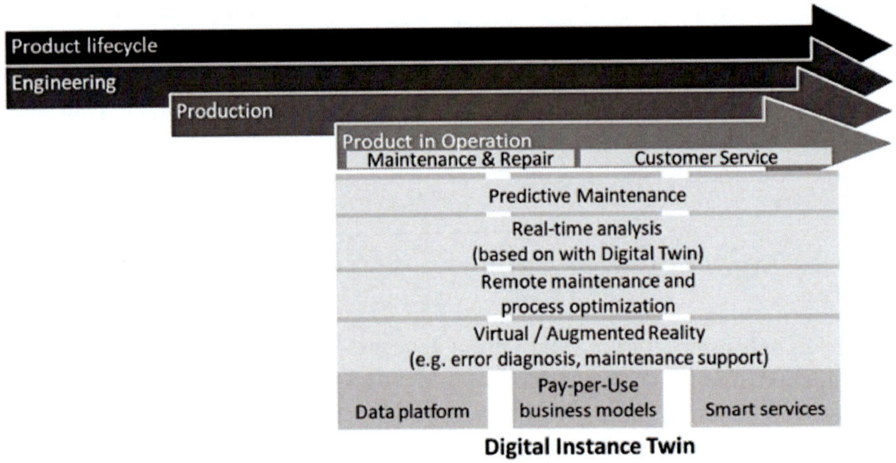

Fig. 3.4 Digital instance twin [11]

physical component. This Mixed Reality Vision can be used for inspection purposes. Any inspection related information can be directly attached to the component. to maintain the continued airworthiness of the aircraft and prevent a failure [24].

3.4 Main Expressions of Digital Twin in Context of Industry 4.0

In the era of Industry 4.0, the question arises as to how Digital Twin is positioned for this industrial initiative. Within Industry 4.0 Reference Architecture Model (RAMI 4.0) [41], Digital Twin is primarily assigned to the Lifecycle & Value stream axis in order to provide the horizontal integration within the product lifecycle (Fig. 3.5) as described in Sect. 3.2. Digital Twin is literally floating in the solution space of RAMI 4.0 because the assignment to the axes Hierarchy Level and Layer is much less clear [41]. Basically, Digital Twin remains related to Product and Asset in context of Industry 4.0. In this sense, a precise description of Digital Twin remains difficult [42].

In order to better classify Digital Twin, it must be clarified which of the following 8 hypotheses apply [42]:

1. Digital Twin is a digital representation of an asset.
2. Digital Twin is in several places at the same time.
3. Digital Twin has a variety of states.
4. In an interaction situation, Digital Twin has a context-specific state.

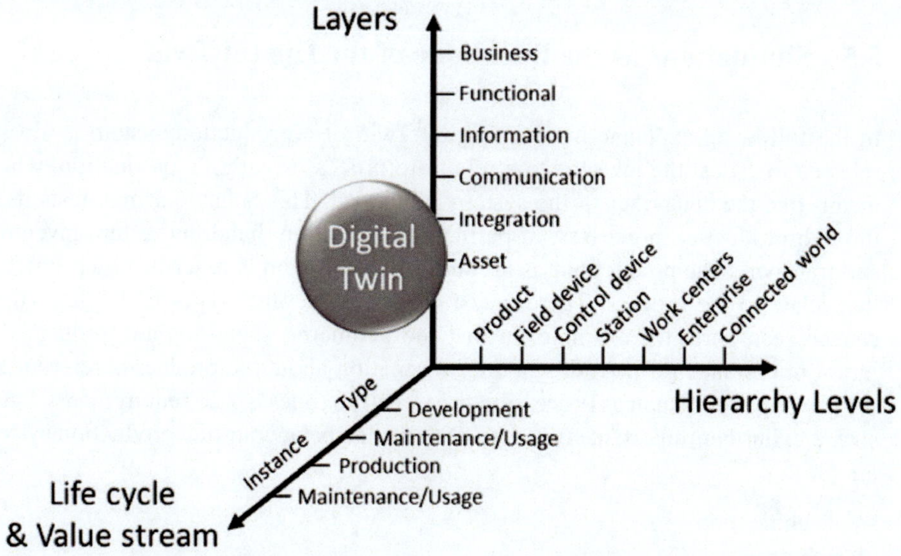

Fig. 3.5 Digital Twin in context of RAMI 4.0 [41]

5. The information model for Digital Twin is infinitely large, it is a real information model.
6. The real information model can finitely be approximated for a specific application scenario and thus becomes a rational information model.
7. The rational information model cannot be stored in one place.
8. The rational information model is never fully visible.

An asset is an object of value. What an asset is specifically for a specific application scenario depends on the application scenario. Whether this object is tangible or intangible, a product or a production system, a type or an instance, is irrelevant. Digital Twin becomes visible at several locations at the same time along the product lifecycle and interacts with an actuator (human, machine, etc.) at these locations [43]. As an outcome, Digital Twin has a variety of states depending on locations and timeline. Digital Twin is put into a context-specific state in a concrete interaction situation. In order to be able to interact with Digital Twin in a specific application scenario, an approximate information model must exist. We refer to this, again based on mathematics, as the rational information model. As can be seen in Fig. 3.1, the data of the rational information model is distributed along the product lifecycle. They are not stored in one container, for example in a central database. In order to feed the data required for a specific interaction situation to a specific actuator, this data must be transported via a suitable interface infrastructure. Consequently, all data of the rational information model are never completely visible [42]. Therefore, an adequate specialization and conscious reduction of Digital Twin to specific application cases remains inevitable, especially if a desired, limited implementation period is taken into account [44].

3.5 Simulations as the Backbone of the Digital Twin

In the following explanations, the Digital Twin of a production system is always referred to. Thus, the investigation relates to processes within a production, which means that the behaviour of the system is mapped. This behaviour manifests itself through production processes, in particular production, handling, setup, inventory and transport. The product that is produced in the system is described by properties that relate to the process. Therefore, a difference to other types of Digital Twins can be seen here, for example, in that the geometric shape of the product is of minor importance [45]. At this point, information about the product is relevant for the investigations, such as processing times, set-up times, space requirements, batch size, etc. Further important entities to describe the behaviour of a production system are [46]:

- Machine tools
- Robots
- Workers
- Workstations

– Storage and buffers
– Paths and transport routes
– Transport devices
– Conveyors
– Load carriers

This is a superordinate classification of the relevant parameters. The various parameters that are further relevant for the Digital Twin of a production system are discussed in Chap. 9 of this book.

Simulations are therefore used to reproduce the behaviour of a production system in the virtual environment. The simulations thus form the backbone for the Digital Twin. The definition of simulation is the reproduction of a dynamic process in a system by means of an experimental model in order to gain knowledge that can be transferred to reality [47]. Thereby, the model represents the object and the system in a simplified form and under consideration of a concrete purpose [48].

When applied to production processes in industrial companies, the real production system is transferred into a simplified model, which can be used to simulate the dynamic behaviour and complex interactions of the production system [49]. Simulation of a production system is thus a key component of the digital model, Digital Shadow and Digital Twin, if they are supposed to represent a production system virtually.

More precisely specified, discrete event simulation methods, also called Discrete Event Simulation (DES), are usually used. In this method, system states are changed at certain points in time, so that the sequence of this simulation is defined. This form of simulation is common for simulations of production systems, which results in a relatively low calculation effort [48]. Figure 3.6 illustrates how the simulation supports Digital Twin as well as Digital Shadow and digital Model. Depending on which technology of the three is applied, a different exchange with the physical production system is performed. With the digital model, the model is built up once or adjusted as required and selectively. With the help of the simulation, analyses

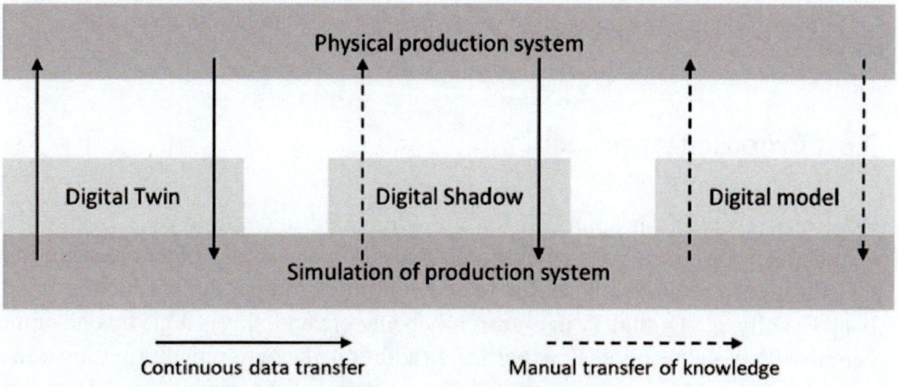

Fig. 3.6 Simulation for Digital Twin, Digital Shadow and Digital Model

are executed, which in turn are applied as findings for the real model. The digital model can also be seen as a synonym for DES at this point. Unlike the digital model, the Digital Shadow is continuously supplied with up-to-date data from the physical production system [50]. Accordingly, a data link is required, but this is only one-way from the real system to the simulation model. In contrast, the Digital Twin extends this concept by a further data connection to the real system [51].

Many fields of application exist today for simulation models of a production system, e.g. planning of factory plants, layout optimization in the shop floor, approval processes in the area of reconstruction and fire protection, optimization of production processes and material transport. A detailed description of the applications is given in the book in Chap. 9. According to Wenzel et al. [52], simulation in particular is a core element of the digital factory and is becoming increasingly important as a result of developments in the area of digitization. Simulation in production and logistics has been scientifically investigated and established for a long time (e.g. [53–55]). According to Nyhuis et al. [56], their use supports companies in optimising logistical targets, e.g. with regard to adherence to schedules, throughput times, performance, inventory and costs. The benefits in the area of material flow planning are rated by companies as high or very high [57].

Nevertheless, current studies prove that the use of simulation models for production systems in small and medium-sized enterprises is still not standard [58]. The main reasons for this backlog are the following organizational and technical obstacles [58–60]:

1. Non-transparent procurement costs (e. g. due to manual or inefficient creation of Digital Twin).
2. Required IT expertise too high (e. g. due to inefficient or overly expensive services).
3. Unpredictable operating expenses (e. g. due to manual or inefficient adaptation of Digital Twin).
4. Lack of knowledge regarding available simulation tools and application areas as well as the achievable benefits.

In the following an approach to overcome these obstacles is presented.

3.6 Proposed Approach

With this research, an approach for the automated generation of a Digital Twin in manufacturing in a built environment based on scans and object recognition is investigated to provide a basis for various optimizations in the production process [61]. Usually, if a Digital Twin is currently being created, this is done by one or more people with the appropriate IT expertise. In addition, the required information is made available to these persons, for example layout plans of the production hall, machine lists, product data and so on. This process requires a high manual effort, which among other things results from the fact that specific knowledge and information

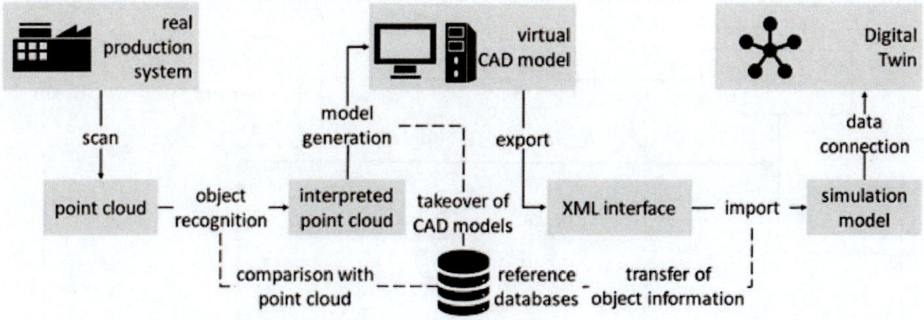

Fig. 3.7 Proposed approach for Digital Manufacturing Twin in the built environment

from production must be transferred to the area of software development. During this domain transition, increased effort is required for coordination, transformation and error prevention.

Figure 3.7 illustrates the functional scheme of the novel approach. In general, the generation of the Digital Twin is shortened and simplified by distinguishing between three parameter groups. With the distinction into three groups, the structure described above is broken down by implementing an efficient data acquisition procedure for each parameter group. With this procedure it is possible to automate the creation of the Digital Twin in many places, to standardize the transformation of the information, and thus to obtain an efficient and cost-effective creation process [62].

The first parameter group consists of spatial information, which is recorded by a scan. For the scanning process, several procedures, such as photogrammetry and laser scanning, are generally conceivable, depending on specific characteristics of the production system and the building [63]. A detailed point cloud is generated by the scanning [64]. This point cloud is the basis for the next steps and at the same time the overall layout of the hall and the positions of the individual objects in the digital model are created [65]. The details about the scanning process are explained in the fourth chapter of this book.

The second parameter group contains the objects of production. The procedure to collect the required information is based on the scan, where object recognition identifies the entities of a production, such as machines or routes [66]. The objective is to recognize all visible objects of the production and to include them automatically into the digital model. For object recognition, methods of artificial intelligence are utilized [67]. All details about this process are explained in Chaps. 5–8.

Finally, the third group is all company-specific information that describes the production processes, such as linkages, sequencing, buffer and so on. This information cannot be automatically acquired via a scan and is usually different for every company. This expert knowledge and the company-specific information of the production system need to be acquired by forms or expert interviews and inserted in the simulation modelling process [68]. It is also conceivable that data could be taken directly from IT systems. Further descriptions of the third parameter group can be found in Chap. 9.

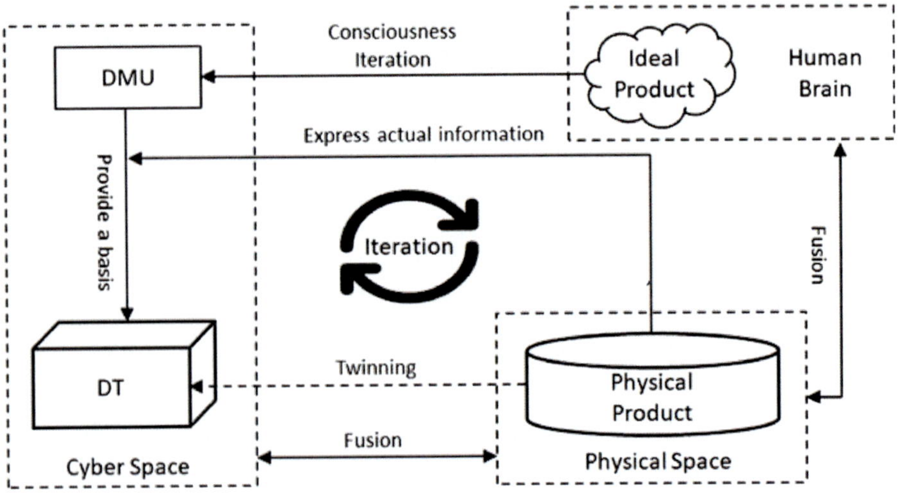

Fig. 3.8 The relationship between DMU and Digital Twin [70]

By using the taxonomy as presented in Sect. 3.2, the following classification of this approach can be provided. With regard to data link, this Digital Twin is uni-directional. Dimension purpose is for data transfer und repository. Dimension accuracy is partial (e. g., adjustable according to the process requirements). Dimension interface is machine-to-machine. Synchronization between the physical and the digital part is not existent. Data input is fed with raw data. Time of creation id pre-defined by the built environment: the physical part first.

Apart of the use of the related taxonomy, a differentiation from the term Digital Mock-up (DMU) is useful [69]. In summary, DMU and Digital Twin, as two aspects that define ideal products and physical products, can be organically unified, as shown in Fig. 3.8. The merge of DMU and Digital Twin reflects the high integration of cyberspace, physical space, and consciousness space. In this view, DMU can be understood as the basis or the previous stage of Digital Twin. Therefore, Digital Twins will become the"middleware" for in-depth communication between human and machine [70].

3.7 Summary and Further Research

The Digital Twin offers great potential for manufacturing companies [71]. Research in this field has increased rapidly in recent years and more and more applications can be found in practice [72]. Nevertheless, the technology of the Digital Twin does not yet infuse the entire industry due to the lack of applications for specific scenarios and difficult implementation [73].

In this chapter, the different types of Digital Twins in manufacturing companies were first described. An important distinction here is that in the following the

Digital Twin is addressed by production systems: Digital Manufacturing Twin [74]. At this point, the simulation of manufacturing processes is the basis for the investigations. With the simulation, which is connected to the real production system via data interfaces, the technology of the Digital Twin is made available [75]. A particular importance is dedicated the built environment where no or not complete or not exact 3D documentation is available [76].

Since advanced IT expertise is required to generate the Digital Twin and its creation or benefits remain unclear to many companies, the Digital Twin has not yet achieved greater diffusion in industry [77]. To overcome these obstacles, a novel approach to create a Digital Manufacturing Twin in a built environment is presented. Here, a flexible, cost-effective and efficient approach is applied with fast scans of the factory floor, object recognition and a highly automated simulation model construction. The further explanations in this book will go through the details of this approach. As the outcome of this approach, production planner gets a powerful means to optimize the production processes in the built environment. Intermediate results can be used for layout planning, construction progress control and documentation purposes [78].

References

1. Trauer J, Schweigert-Recksiek S, Engel C, Spreitzer K and Zimmermann M (2020) What is a Digital Twin? Definitions and insights from an industrial case study in technical product development. Int Design Conf Design 2020. https://doi.org/10.1017/dsd.2020.15
2. Sieg O, Völl, C (2020) Digital twin and Product Lifecycle Management (PLM) as part of the corporate strategy [Digital Twin und PLM als Teil der Unternehmensstrategie]. ZWF Zeitschrift fuer Wirtschaftlichen Fabrikbetrieb 115:38
3. Stark R, Damerau T (2019) Digital Twin. In: Chatti S, Laperrière L, Reinhart G, Tolio T (eds) The international academy for production engineering. CIRP encyclopedia of production engineering, 2nd edn. Springer, Berlin Heidelberg. https://doi.org/10.1007/978-3-642-35950-7_16870-1
4. Lo CK, Chen CH, Zhong RY (2021) A review of Digital Twin in product design and development. Adv Eng Inform
5. Ríos J, Morate FM, Oliva M, Hernández JC (2016) Framework to support the aircraft digital counterpart concept with an industrial design view. Int J Agile Syst Manage 9(3):212–231. https://doi.org/10.1504/IJASM.2016.079934
6. Ladj A, Wang Z, Meski O, Belkadi F, Ritou M, Da Cunha C (2020) A knowledge-based Digital Shadow for machining industry in a Digital Twin perspective. J Manuf Syst. https://doi.org/10.1016/j.jmsy.2020.07.018
7. Weber U, Hong NA, Schäfer V, Peters P, Vettermann S (2015) Synced factory Twins: next lever digital manufacturing. ProduktDataJournal 2(2015):8–13
8. Biesinger F, Meike D, Kraß B, Weyrich M (2018) A case study for a digital twin of body-in-white production systems—general concept for automated updating of planning projects in the digital factory. In: 23rd international conference on emerging technologies and factory automation (ETFA). IEEE, pp 19–26. https://doi.org/10.1109/ETFA.2018.8502467
9. Donoghue I, Hannola L, Papinniemi J, Mikkola A (2018) The benefits and impact of digital twins in product development phase of PLM. In: Chiabert P, Bouras A, Noël F, Ríos J (eds) PLM 2018: product lifecycle management to support industry 4.0. Springer International Switzerland, pp 432–441. https://doi.org/10.1007/978-3-030-01614-2_40.

10. Stark J (2015) Product lifecycle management, vol 1. Springer International Publishing, Switzerland. https://doi.org/10.1007/978-3-319-17440-2
11. Wagner L (2020) Kein Digital Twin ohne digitale Durchgängigkeit. Zeitschrift für den wirtschaftlichen Fabrikbetrieb, Band 115, April 2020, pp 43–45. https://doi.org/10.3139/104.112315
12. da Silva Serapião Leal G, Guédria W, Panetto H (2019) An ontology for interoperability assessment: a systemic approach. J Indus Inform Integr 16:100100. https://doi.org/10.1016/j.jii.2019.07.001
13. Ruppert T, Abonyi J (2020) Integration of real-time locating systems into digital twins. J Indus Inform Integr 20:100174. https://doi.org/10.1016/j.jii.2020.100174
14. El-khoury J, Berezovskyi A, Nyberg A (2019) An industrial evaluation of data access techniques for the interoperability of engineering software tools. J Indus Inform Integr 15(2019):58–68. https://doi.org/10.1016/j.jii.2019.04.004
15. Haenlein M, Kaplan A, Tan CW, Zhang P (2019) Artificial intelligence (AI) and management analytics. J Manage Anal 6(4):341–343. https://doi.org/10.1080/23270012.2019.1699876
16. Schmitt J, Bönig J, Borggräfe T, Beitinger G, Deuse, J (2020) Predictive model-based quality inspection using machine learning and edge cloud computing. Adv Eng Inform 45:101101. https://doi.org/10.1016/j.aei.2020.101101
17. Li C, Zhang S, Qin Y, Estupinan E (2020) A systematic review of deep transfer learning for machinery fault diagnosis. Neurocomputing 407:121–135. https://doi.org/10.1016/j.neucom.2020.04.045
18. Voell C, Chatterjee P, Rauch A, Golovatchev J (2018) How Digital Twins enable the next level of PLM: a guide for the concept and the implementation in the internet of everything era. In: Chiabert P et al. (eds) Product lifecycle management to support industry 4.0. Springer International, Cham, pp 238–249. https://doi.org/10.1007/978-3-030-01614-2_22.
19. Xu J, Guo T (2021) Application and research on digital twin in electronic cam servo motion control system. Int J Adv Manuf Technol 112:1145–1158. https://doi.org/10.1007/s00170-020-06553-7
20. Martínez GS, Sierla SA, Karhela TA, Lappalainen J, Vyatkin V (2018) Automatic generation of a high-fidelity dynamic thermal-hydraulic process simulation model from a 3D plant model. IEEE Access 6:45217–45232. https://doi.org/10.1109/ACCESS.2018.2865206
21. Gehrke I, Schauss M, Küsters D, Gries T (2020) Experiencing the potential of closed-loop PLM systems enabled by industrial internet of things. Procedia Manuf 45:177–182. https://doi.org/10.1016/j.promfg.2020.04.091
22. Biahmou A, Emmer C, Pfouga A, Stjepandić J (2016) Digital master as an enabler for industry 4.0. Adv Transdisciplinary Eng 4:672–681. https://doi.org/10.3233/978-1-61499-703-0-672
23. Sommer M, Stjepandić J, Stobrawa S, von Soden M (2019) Automatic generation of digital twin based on scanning and object recognition. Adv Transdisciplinary Eng 10(2019):645–654. https://doi.org/10.3233/ATDE190174
24. Utzig S, Kaps R, Azeem SM, Gerndt A (2019) Augmented reality for remote collaboration in aircraft maintenance tasks. 2019 IEEE aerospace conference. https://doi.org/10.1109/AERO.2019.8742228
25. Fukuda S, Lulić Z, Stjepandić J (2013) FDMU-functional spatial experience beyond DMU? In: Proceedings of the 20th ISPE international conference on concurrent engineering, CE 2013. IOS Press, Amsterdam, 2013, pp 431–440. https://doi.org/10.3233/978-1-61499-302-5-431
26. Yao J, Zhang Z (2020) C3+ATO function simulation and verification analysis based on timed automata. Adv Transdisciplinary Eng 14:166–173. https://doi.org/10.3233/ATDE200229
27. Peng Y, Zhao S, Wang H (2021) A digital twin based estimation method for health indicators of DC-DC converters. IEEE Trans Power Electron 36(2):9141430. https://doi.org/10.1109/TPEL.2020.3009600
28. Besbes M, Zolghadri M, Costa Affonso R, Masmoudi F, Haddar M (2020) 3D facility layout problem. J Intell Manuf. https://doi.org/10.1007/s10845-020-01603-z
29. Usuga Cadavid JP, Lamouri S, Grabot B, Pellerin R, Fortin A (2020) Machine learning applied in production planning and control: a state-of-the-art in the era of industry 4.0. J Intell Manuf. https://doi.org/10.1007/s10845-019-01531-7

30. Donhauser T, Endter U, Schmidt T, Baier L, Schuderer P, Franke J (2020) Real-time image of production. Material flow simulation during operation | [Betriebsbegleitende materialflusssimulationechtzeitabbild der produktion]. WT Werkstattstechnik 110(4):214–219
31. He B, Cao X, Hua Y (2021) Data fusion-based sustainable digital twin system of intelligent detection robotics. J Cleaner Prod 280:124181. https://doi.org/10.1016/j.jclepro.2020.124181
32. Uhlenbrock L, Jensch C, Tegtmeier M, Strube J (2020) Digital twin for extraction process design and operation. Processes 8(7):866. https://doi.org/10.3390/pr8070866
33. Wallis R, Stjepandic J, Rulhoff S, Stromberger F, Deuse J (2014) Intelligent utilization of digital manufacturing data in modern product emergence processes. Adv Transdisciplinary Eng 261–270. https://doi.org/10.3233/978-1-61499-440-4-261
34. Lu Y, Xu X (2019) Cloud-based manufacturing equipment and big data analytics to enable on-demand manufacturing services. Robot Comput Integr Manuf 57:92–102. https://doi.org/10.1016/j.rcim.2018.11.006
35. Peruzzini M, Stjepandić J (2018) Editorial to the special issue "Transdisciplinary analytics in supply chain management." J Manage Anal 5(2):75–80. https://doi.org/10.1080/23270012.2018.1443405
36. Barrios P, Eynard B, Danjou C (2018) Towards a digital thread between industrial internet of things and product lifecycle management: experimental work for prototype implementation. In: Chiabert P et al. (eds) Product lifecycle management to support industry 4.0. Springer International, Cham, pp 273–282. https://doi.org/10.1007/978-3-030-42250-9_26
37. Barata J, Pereira V, Coelho M (2020) Product biography information system: a lifecycle approach to Digital Twins. IEEE Trans Syst Man Cybern Syst 899–904. https://doi.org/10.1109/SMC42975.2020.9283061
38. van der Valk H, Haße H, Möller F, Arbter M, Henning JL, Otto B (2020) A taxonomy of Digital Twins. In: Anderson B, Thatcher J, Meservy R (eds) Proceedings of the 26th Americas conference on information systems, pp 1–10
39. Bicocchi N, Cabri G, Mandreoli F, Mecella M (2019) Dynamic digital factories for agile supply chains: an architectural approach. J Indus Inform Integr 15:111–121. https://doi.org/10.1016/j.jii.2019.02.001
40. Xu Y, Päivärinta T, Kuvaja P (2020) Digital Twins as software and service development ecosystems in industry 4.0: towards a research agenda. Communications in computer and information science, 1210 CCIS, pp 51–64. https://doi.org/10.1007/978-981-15-7530-3_5
41. DIN SPEC 91345 (2016) Referenzarchitekturmodell Industrie 4.0 (RAMI4.0). Deutsches Institut für Normung (DIN) e.V. https://doi.org/10.31030/2436156
42. Deuter A, Pethig F (2019) The Digital Twin theory—Eine neue Sicht auf ein Modewort. Industrie 4.0 Management 35(1):27–30. https://doi.org/10.30844/I40M_19-1_S27-30
43. Ma X, Tao F, Zhang M, Wang T, Zuo Y (2019) Digital twin enhanced human-machine interaction in product lifecycle. Procedia CIRP 83:789–793. https://doi.org/10.1016/j.procir.2019.04.330
44. Adamenko D, Kunnen S, Pluhnau R, Loibl A, Nagarajah A (2020) Review and comparison of the methods of designing the Digital Twin. Procedia CIRP 91:27–32. https://doi.org/10.1016/j.procir.2020.02.146
45. Lu Q, Chen L, Li S, Pitta M (2020) Semi-automatic geometric digital twinning for existing buildings based on images and CAD drawings. Autom Construct 115:103183. https://doi.org/10.1016/j.autcon.2020.103183
46. Denkena B, Dittrich M-A, Stobrawa S, Stjepandic J (2019) Automated generation of a digital twin using scan and object detection for data acquisition, 18. ASIM Fachtagung: Simulation in Produktion Und Logistik, Wissenschaftliche Scripten, Auerbach, pp 49–58
47. VDI-Richtline 3633, Blatt 1 (2014) Simulation von Logistik-, Materialfluss- und Produktionssystemen - Grundlagen. Beuth, Berlin
48. Gutenschwager K, Rabe M, Spieckermann S, Wenzel S (2017) Simulation in Produktion und Logistik - Grundlagen und Anwendungen. Springer Vieweg, Wolfenbüttel,. https://doi.org/10.1007/978-3-662-55745-7

49. Emmer C, Fröhlich A, Stjepandić J (2013) Advanced engineering visualization with standardized 3D formats, IFIP advances in information and communication technology, vol 409. Springer, Berlin Heidelberg, pp 584–595. https://doi.org/10.1007/978-3-642-41501-2_58
50. Pfouga A, Stjepandić J (2018) Leveraging 3D geometric knowledge in the product lifecycle based on industrial standards. J Comput Des Eng 5(1):54–67. https://doi.org/10.1016/j.jcde.2017.11.002
51. Ktari A, El Mansori M (2020) Digital twin of functional gating system in 3D printed molds for sand casting using a neural network. J Intell Manuf. https://doi.org/10.1007/s10845-020-01699-3
52. Wenzel S, Peter T (2017) Prozess-Simulation in kleinen und mittleren Unternehmen mittels des Bausteinkastensystems KMUSimMetall. Simul Produktion und Logistik 2017:433–442
53. Spieckermann S (2005) Diskrete, ereignisorientierte Simulation in Produktion und Logistik - Herausforderungen und Trends. Simulation und Visualisierung 2005 - SimVis. SCS Publishing House, Erlangen, pp 3–14
54. Rabe M, Spieckermann S, Wenzel S (2008) Verifikation und Validierung für die Simulation in Produktion und Logistik: Vorgehensmodelle und Techniken. Springer, Berlin,. https://doi.org/10.1007/978-3-540-35282-2
55. Wenzel S, Weiß M, Collisi-Böhmer S, Pitsch H, Rose O (2007) Qualitätskriterien für die Simulation in Produktion und Logistik: Planung und Durchführung von Simulationsstudien. Springer-Verlag. https://doi.org/10.1007/978-3-540-35276-1
56. Nyhuis P, Wiendahl HP (2012) Logistische Kennlinien: Grundlagen, Werkzeuge und Anwendungen. Springer, Berlin. https://doi.org/10.1007/978-3-540-92839-3
57. Bierschenk S, Ritter A, Kuhlmann T (2005) Stand der Digitalen Fabrik bei kleinen und mittelständischen Unternehmen. Fraunhofer-IRB-Verlag
58. Bischoff J et al. (2015) Erschließung der Potenziale der Anwendungen von, „Industrie 4.0" im Mittelstand. Studie im Auftrag des Bundesministeriums für Wirtschaft und Energie (BMWi). Mülheim an der Ruhr
59. Denkena B, Winter F, Hauck S (2016) Anlernen von ablauforientierten simulationsmodellen. Productivity 21(3):52–54
60. Weissman A, Wegerer S (2019) Unternehmen 4.0: Wie Digitalisierung Unternehmen & Management verändert. In: Erner M (eds) Management 4.0—Unternehmensführung im digitalen Zeitalter. Springer-Verlag, Berlin, pp 43–78. https://doi.org/10.1007/978-3-662-57963-3_2
61. Sommer M, Stjepandić J, Stobrawa S, von Soden M (2021) Automated generation of a Digital Twin in manufacturing for a built environment using scan and object detection. J Indus Inform Integr (in press)
62. Denkena B, Dittrich MA, Stobrawa S, Stjepandić J (2020) Efficient generation of a digital twin using object detection for data acquisition and XML-interface for model creation. Procedia CIRP 93:274–279. https://doi.org/10.1016/j.procir.2020.03.016
63. Moon D, Chung S, Kwon S, Seo J, Shin J (2019) Comparison and utilization of point cloud generated from photogrammetry and laser scanning: 3D world model for smart heavy equipment planning. Autom Construct 98:322–331. https://doi.org/10.1016/j.autcon.2018.07.020
64. Wang K, Kim M-K (2019) Applications of 3D point cloud data in the construction industry: a fifteen-year review from 2004 to 2018. Adv Eng Inform 39:306–319. https://doi.org/10.1016/j.aei.2019.02.007
65. Ma Z, Liu S (2018) A review of 3D reconstruction techniques in civil engineering and their applications. Adv Eng Inform 37:163–174. https://doi.org/10.1016/j.aei.2018.05.005
66. Bondar S, Salem B, Stjepandić J (2018) Indoor object reconstruction based on acquisition by low-cost devices. Adv Transdisciplinary Eng 7(2018):113–122. https://doi.org/10.3233/978-1-61499-898-3-113
67. Salem B, Stjepandić J, Stobrawa S (2019) Assessment of methods for industrial indoor object recognition. Adv Transdisciplinary Eng 10:390–399. https://doi.org/10.3233/ATDE190145

68. Denkena B, Stobrawa S, Sommer M, Stjepandic J, Von Soden M (2020) Production planning using the digital twin | [Effiziente Erstellung eines digitalen Zwillings zur Verwendung im Produktionsmanagement; Produktionsplanung mit dem digitalen Zwilling]. WT Werkstattstechnik 110(10):661–665
69. Dai S, Zhao G, Yu Y, Bao Q (2020) The evolution from digital mock-up to Digital Twin. Front Artif Intell Appl Mach Learn Artif Intell 332:184–191. https://doi.org/10.3233/FAIA200781
70. Riascos R, Levy L, Stjepandić J, Fröhlich A (2015) Digital mock-up. In: Stjepandic J et al. (eds) Concurrent engineering in the 21st century: foundations, developments and challenges, Springer International Switzerland, pp 355–388
71. Schleich B, Dittrich MA, Clausmeyer T, Damgrave R, Erkoyuncu JA, Haefner B, de Lange J, Plakhotnik D, Scheidel W, Wuest T (2019) Shifting value stream patterns along the product lifecycle with digital twins. Procedia CIRP 86:3–11. https://doi.org/10.1016/j.procir.2020.01.049
72. Redelinghuys AJH, Basson AH, Kruger K (2019) A six-layer architecture for the digital twin: a manufacturing case study implementation. J Intell Manuf. https://doi.org/10.1007/s10845-019-01516-6
73. Hong Lim KY, Zheng P, Chen CH (2019) A state-of-the-art survey of Digital Twin: techniques, engineering product lifecycle management and business innovation perspectives. J Intell Manuf. https://doi.org/10.1007/s10845-019-01512-w
74. Polini W, Corrado A (2021) Digital twin of stone sawing processes. Int J Adv Manuf Technol 112:121–131. https://doi.org/10.1007/s00170-020-06384-6
75. Zhong RY, Ge W (2018) Internet of things enabled manufacturing: a review. Int J Agile Syst Manage 11(2):126–154. https://doi.org/10.1504/IJASM.2018.092545
76. Guo H, Zhu Y, Zhang Y, Ren Y, Chen M, Zhang R (2021) A digital twin-based layout optimization method for discrete manufacturing workshop. Int J Adv Manuf Technol 112:1307–1318. https://doi.org/10.1007/s00170-020-06568-0
77. Kritzinger W, Karner M, Traar G, Henjes J, Sihn W (2018) Digital Twin in manufacturing: a categorical literature review and classification. IFAC-PapersOnLine 51(11):1016–1022. https://doi.org/10.1016/j.ifacol.2018.08.474
78. Otero R, Lagüel S, Garrido I, Arias P (2020) Mobile indoor mapping technologies: a review. Autom Construct 120:103399. https://doi.org/10.1016/j.autcon.2020.103399

Chapter 4
Scan Methods and Tools for Reconstruction of Built Environments as Basis for Digital Twins

Markus Sommer and Klaus Seiffert

4.1 Introduction

In general, a 3D scan describes the measurement of an object or an environment depicting its shape and appearance. Its result is a digital reconstruction of the surveyed scene. Ideally, the real object is completely depicted within the model by a cloud of 3D data points containing the position and color information. The necessity for a new scan arises, if a structure is not yet available as a digital twin or if the existing model no longer corresponds to the reality due to alterations [1].

When scanning a 3D environment, two different approaches are to be distinguished:

Active Technologies
With these technologies the surface of an object or a complete environment is captured by active methods, either mechanically or by illuminating them using an additional source of radiation. A 3D profile is generated, as seen from the perspective from the point of the measurement. All active methods interact with the object to be scanned. In the case of large-scale objects tactile methods are too slow to be feasible. Radiometric measurements include moving light sources, the usage of structured or colored visible light sources, time of flight and phase-shift detecting laser-based methods. All these approaches illuminate the surface of the environment by an artificial source of light and use the reflection to calculate the distance between a sensor and an object.

M. Sommer (✉) · K. Seiffert
isb innovative software businesses GmbH, Otto-Lilienthal-Strasse 2, 88046 Friedrichshafen, Germany
e-mail: sommer@isb-fn.de

K. Seiffert
e-mail: seiffert@isb-fn.de

J. Stjepandić et al. (eds.), *DigiTwin: An Approach for Production Process Optimization in a Built Environment*, Springer Series in Advanced Manufacturing,
https://doi.org/10.1007/978-3-030-77539-1_4

Passive Technologies

These technologies do not actively disturb the object to be scanned during the measurement. Instead, a sensor picks up the reflected or emitted radiation from the surface of an object in its normal environment. Through interpreting the depicted scene, the 3D structure is inferred. Typically, the sensor of a camera captures visible light, meaning common photographs or videos are being used. To reconstruct the complete structure of an object or sounding a series of many images from many different perspectives is required. In contrast to active technologies, passive technologies cover a wider range of applications [2].

Technical Development

Passive methods are as old as photography itself. The theory of photogrammetry was developed parallel to photography in the middle of the nineteenth century. Firstly, it was used to survey buildings. With the invention of the stereo camera in 1907, automated contour lines of landscapes could be generated for the first time. In 1930 bundle block adjustment was described as a theory, a procedure to optimize calculations of depths when using batches of images and with the introduction of computers starting around 1960, its implementations became very efficient.

Around 1950 the U.S. military started to experiment with the first optical measuring devices and in Scotland an electronic probing machine for precise measurements was developed. The two measuring devices are forming the technological basis for today's laser scanners. The laser-based LIDAR system for investigating topographic scenarios was developed by the U.S. military. After the disclosure of the technology by the Army, 3D acquisition of structures in civilian applications became much more efficient. In 1990, for the first time a 3D laser scanner was developed explicitly for industrial use. At the time, this laser scanner was fast, accurate and cost effective. Surveying technicians and civil engineers could now easily and efficiently create 3D images. However, in the first step, generated point clouds did not contain any information about color [3].

With the emerging and progression of digital photography photogrammetric systems have become far easier to implement. As the resolution in digital photography has increased continuously to the point where it is sufficient to differentiate between small individual features in images, photogrammetric systems have seen a rise in popularity. In recent years, even standard cameras systems achieve a high enough resolution and simple setups like using build-in cell phone are tested for their potential applications. Parallel the capability of drone systems are constantly developing and some autonomous measurement systems have already been implemented. However, these technologies are mainly tested in research and are not yet implemented in standardized products to perform scans in any environments.

In the cases of surveying and construction, compared to photogrammetry, the laser scanner is much more precise and robust against external influences. However, as soon as larger areas are to be scanned, in the past reference marks had to be positioned within the scan. Through matching these reference marks, the individual scans could be merged with each other. Due to the ever-higher resolution of the laser scanners, the reference marks are no longer necessary in all cases. Additionally, after the scanning process of the laser, photographic images are taken and the point cloud is enriched with color information.

Advantages and Disadvantages

Both, active and passive technologies have their advantages and disadvantages. When using an active technology, the surface of an object is scanned. The incoming beam can be reflected in differently depending on the surface and angle, for instance with laser scanners the laser beam might pass through a sheet of glass and the transparent material cannot be detected. Also, reflective surfaces such as chrome-plated materials can cause problems. Depending on the type of scanning signal, the maximum distance is limited. Scanners, measuring by the time-of-flight principle, have a high accuracy up to 40 m. Today, modern terrestrial laser scanners can measure up to 200 m without any problems. Scanning with a laser scanner still requires expert knowledge [4].

The passive technology has the great advantage of not being potentially damaging to any object, especially compared to mechanical scanning methods. The on-site recording time is less than in the active case by a factor of three. The resolution of the scan can be easily adjusted, the more images from different angles of the scene are taken, the higher the resulting resolution of the scan. A large disadvantage of the passive technology is, that the object to be scanned needs to have distinctive and prominent features visible over all individual images to correctly asses the connections in between them. Changing lighting conditions during the measurement have a negative effect on the reconstruction. Homogeneous, flat surfaces and surfaces reflecting sunlight can hardly be reconstructed at all, often resulting holes within the point cloud [5].

Area of Application Regarding the Different Technologies

As soon as an accuracy of a few millimeters is a requirement when measuring a large scene, a laser based scanning technology (an active technology) will almost always be the definite choice. But, high accuracy has a price, the scan is time-consuming and the scanning equipment is expensive.

However, as soon as a resolution within the range of centimeters is sufficient, the use of photogrammetry (passive technology) is likely to be preferred. The investment for the scanning equipment being a camera is smaller by a high factor. The actual measurement can also be performed by non-specialists after a short instruction. Table 4.1 shows some of the technologies and their areas of application [6]:

In this chapter, approaches for acquisition of physical objects in the built environment to be used for generation of the Digital Twin are presented. In Sect. 4.2, requirements for acquisition of shape and position of physical objects in a factory are discussed. In Sect. 4.3, main acquisition devices are presented in a comparative view, followed in Sect. 4.4 by discussion of methodical frameworks for structural investigation of unknown objects. In Sect. 4.5, the approaches for image-based 3D reconstruction are explored. Approaches for 3D reconstruction in production environments are presented in Sect. 4.6. The chapter ends with a summary and future challenges.

Table 4.1 Application different technologies

Technologie	Area of application
Lasers	• Indoor and outdoor application with a range to 200 m • High accuracy with 1 mm
Time of Flight	• Gameing engines • Autmotive Applikation e.g. Passanger detection • HumanMachine Interface • Medicine
Structed Light	• High precision measurement of components
Photogrammetry	• Indoor and outdoor application with a Range to 100 m • Accuracy with 0,5 cm

4.2 Requirements for Acquisition of Shape and Position of Physical Objects in a Factory

In case of a factory, specific physical constraints must be considered which can impair the outcome of the object acquisition: occlusion, dust, dirt, darkness. Therefore, the process of acquisition of shape and position of objects in a factory hall must be carefully planned based on specific requirements. It also includes the selection of the appropriate data acquisition device. The scanning of production facilities is subject to certain requirements being established using the following example from practice.

An exemplary floorplan of a factory is depicted in Fig. 4.1.

Fig. 4.1 Production floorplan

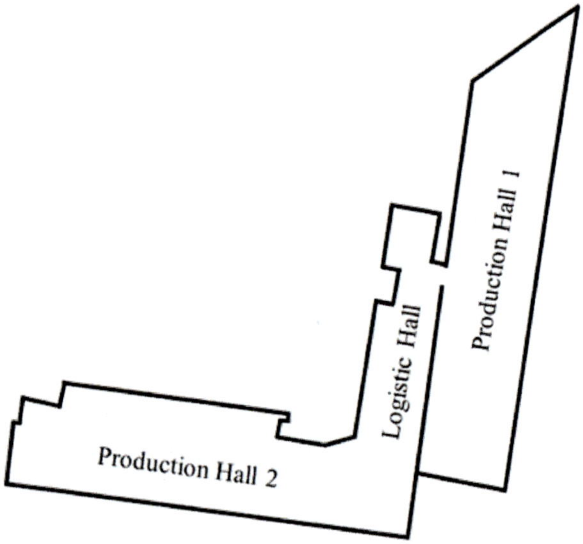

4.2.1 Requirements for the Scan of Manufacturing Facilities

Size of the Facilities

Typical lengths and widths of manufacturing halls are in the range of 20 to 30 m and above. Heights vary from four meter onwards. Production facilities can be divided into several halls separated by doors or rolling gates. Machines in production are typically large objects and can span over several floors resulting in equally sized halls.

Objects Within Production Facilities

Objects to be produced very greatly and shape and size. Consequently, machines range from individual small milling machines to complete machining centers, with the latter being several meters in length, width and height. In many cases the machines are located so close to each other that there are only a small passage of less than one meter of space between them. Then again, buffer zones of several meters often are often placed in front of machines. Additionally, there are workstations consisting of tables and chairs. Robots are found within and even without protective enclosures. In summary, objects within manufacturing facilities are manifold, thus resulting in a challenge to capture large structures and small details at the same time [7].

4.2.2 Requirements from a Production Point of View

Time

Ideally, a scan must be performed without interrupting production. But it is necessary for personal to leave their workplaces for them to not show up in the scan. A maximum time of absence for any employees in the range a few minutes might be tolerable without having to halt production, generally quicker is better.

Execution

It is preferred if the execution of a scan does not require assistance of external specialists. Instead, at best user introduction can be done by a short video tutorial or simple guidelines. The data should be easy to handle and not require specialized hardware components. Ideally, the scan can be performed with freely available products like a camera, drone or alike by employees within the company.

Accuracy

Deriving from resolution and precision, it would be easy to argue higher is better. But the amount of data grows by the power of two considering image resolution and for the subsequent data processing steps this can worse. To distinguish objects within the scan, an accuracy of one to two centimeters at a distance of ten meters is perfectly adequate.

Data Processing
Data is usually stored as raw input while scanning to allow for portable and robust scanning equipment. After the scan, the data is transferred to a processing system for calculation of the three-dimensional point cloud. Algorithms can be provided as a cloud service with standard features like upload, custom settings, progress information and download links for the completed point cloud. The protection of trade secrets and privacy rights is a must that can easily met by an automated processing pipeline [8–10].

4.3 Acquisition Approaches in a Comparative View

The two available methods for creating large scale digital reconstruction of production facilities are laser scanners and photogrammetry. These are described and assessed on meeting the previously established criteria.

4.3.1 Laser Scanner

A laser diode emits coherent electromagnetic radiation. This laser signal is reflected by objects according to their surface properties. A detector receives the signal and evaluates it, different characteristics of laser light can be used for signal interpretation. In the context of virtual three-dimensional reconstruction two main procedures can be distinguished as follows [11–15].

LIDAR
A laser scanner for three-dimensional reconstruction consists of transmitter and receiver. The transmitter emits an extremely short pulse of light and the duration between sent pulse and received signal is measured in all spatial directions. LIDAR is an acronym that can be interpreted as laser imaging, detection, and ranging. LIDAR technology itself has a wide range of applications. Measurement accuracy depends on the precision with which the propagation speed of the light is known within the medium between laser scanner and the object of interest.

Phase Detection
Laser as a coherent source of light emits electromagnetic radiation of constant frequency and waveform. The phase of the emitted light is measured against the phase of the light reflected by objects. By comparison of the phase shift in the continuous harmonic wave of sent and received light, the distance to the measured object can be inferred. Accuracy considerations are the same as in the LIDAR case. For the required accuracy in three-dimensional object reconstruction within manufacturing laser-based measurement methods are highly accurate.

4.3.2 Photogrammetry

Photogrammetry uses generated images, mainly photos or individual frames of videos for the calculation of the position, size and shape of objects. Thus, the measurement of an object is indirectly calculated by analyzing images. The acquisition of geometries in the manufacturing context through the photogrammetric method is realized in two stages.

In the first stage, the object is photographed on site from different angles. Photogrammetry describes the process of retrieving information of one or more objects shown in pictures in an objective manor. Photos stores information with very high density and can be evaluated at any time focusing on different aspects. A photograph is modeled on human vision, which interprets the color and brightness information seen in photos. A temporal separation between recording and evaluation also makes it possible to reconstruct destroyed or already perished buildings three-dimensionally on the basis of historical images. For the recording any systems like traditional cameras, 360-degree cameras, drones with cameras, smartphones or alike can be used. The level of detail and accuracy of the photogrammetric calculation is largely dependent on the resolution and scale of the underlying images.

In the second stage, object features are extracted from the photos. To calculate geometry information from photos, the mapping rules between all photos must be known. The mapping rules are calculated based on the detection of unique object features in the individual photos. When capturing objects with a surface not containing any structures, pure photogrammetry often has a problem to provide a complete reconstruction.

Through the mapping rules the location and direction of the photos relative to each other is well defined. On the basis of this information a sparse point cloud is created. This thin cloud consists largely of the individual feature points used to calculate the position of the photographs relative to each other in space. In a further step, depth maps are computed from overlapping photos using different algorithms. These from the individual photos derived depth maps form the basis for the calculation of a dense point cloud. Using the original information from the photos, all points in the depth map are provided with color values. Then, using the known position of all photos in space, the individual depth maps are combined to form an overall dense point cloud [16, 17].

4.3.3 Comparison of the Different Steps in Generating a Point Cloud by Photogrammetry or Laser Scanner

The following image shows the different steps using photogrammetry and laser scanner to generate a point cloud (Fig. 4.2).

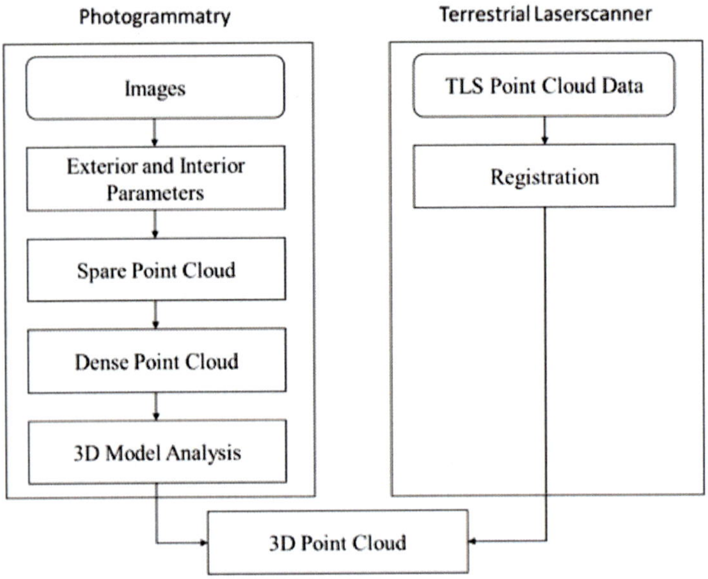

Fig. 4.2 Different steps between photogrammetry and laser scanner

4.3.4 Application of the Different Technologies

In recent years, three dimensional reproductions of production facilities have been mainly generated by laser scanners. Laser scanners have an accuracy in the range of millimeters. A single step in the scanning process with a good resolution considering the height of a ceiling of four meters takes about eight minutes. During these eight minutes, no production can take place at the site. Depending on the density of the machines in production in relation to each other, the total amount of time for a complete scan of a facility of 5000 m^2 can quickly add up to two days.

With photogrammetry the scan is much faster and the production is not necessarily disrupted while taking the images. In a similar setting of 5000 m^2 the time for the recordings amounts to approximately one and a half hours. The calculation of the point cloud based on the photos can be estimated at one to two days. The accuracy of the point cloud is on average about two centimeters on ten meters.

Considering the time for the actual measurement while resulting in models with sufficiently high resolution, photogrammetry-based scan methods are expected to be much more often used in the near future. The technology itself is still a field of major research efforts, further improvements are to be expected. As photogrammetry is in principle much better suited to fulfill the requirements outlined for production environments it is depicted in further details in the following [18].

4.4 SLAM, Structure from Motion, Photogrammetry

Photogrammetry, Structure from Motion (SfM) and Simultaneous localization and mapping (SLAM) are methodological frameworks for the structural investigation of unknown objects. They form the basis for the methods used in building a digital twin of production facilities. All three research disciplines are based on the systematic analysis of sensor data, these being typically images from cameras. Common to all of the methods is the indirect approach to deduce information about the object itself from images, the problem of reconstructing an unknown three-dimensional geometry from two-dimensional data is at the core of all these approaches. Many of the theoretical concepts described in one framework or scientific discipline can therefore also be found in the other two. Even more, whole practical implementations developed in one area of research are directly applied for solving problems in the next.

Still, all these three, photogrammetry, SfM and SLAM are separate disciplines for good reasons. While they are very closely related, the focus of each method is different. All three result in the description of three-dimensional geometries, but it is very different whether to focus on highest, sometimes sub-millimeter precision of a single object or on the reconstruction of large areas with many objects, each of them just having to be identifiable. In the following, a short overview of photogrammetry, SfM and SLAM is given to enable classification, but nevertheless demonstrate the many parallels.

4.4.1 Photogrammetry

Photogrammetry is the methodical analysis, evaluation and interpretation of images in order to obtain reliable information about the objects being depicted. This involves many systematic approaches to determine the exact position and shape of an object by analyzing images. In addition, in parts of photogrammetry, further aspects of the object are inferred through the interpretation of images and attributing classifiers to it.

In photogrammetry, the focus of the investigation is usually one single, clearly delineated object. This object is photographed from different perspectives and a three-dimensional reconstruction of the object is created by evaluating the individual images and linking the information obtained together to one final result. The point cloud, which represents the object under investigation, should be as dense and precise as possible. The aim to obtain a very precise and comprehensive representation of an object can be seen as the main contrast to the following methods SfM and SLAM.

However, photogrammetry is a collection of methods focused on quantifying many different aspects. Thusly, it has developed also an extensive set of tools and models all aimed at achieving high precision in their respective tasks. The goal of photogrammetry is generally to be able to make accurate, but most importantly reliable statements. This includes, for example, checking metric tolerances through image analysis, such as measuring gap dimensions. For this class of statements, it is therefore an elementary component not only to achieve the highest possible quality, but also to be able to quantify measurement inaccuracy, due to the methods and calculations used.

In order to provide reliable, high-precision analyses, photogrammetry begins with the mathematical description of its measuring instruments, for example camera intrinsic and the calibration. Systematic error description and propagation are considered in order to estimate uncertainties. Redundant measurements, such as overlapping images, are used to link the individual pieces of information and to increase precision. Thus, in photogrammetry systematic approaches to quality assurance of all individual steps and the entire process are likely to have a sound theoretical basis and to be well implemented.

Typically, creating a final complete three-dimensional reconstruction from several individual images involves a toolchain containing many steps. Complex non-linear mathematical models can be necessary components in some phases. Individual modules are optimized for themselves and also against each other in order to obtain an overall optimized result.

In summary, photogrammetry is a collection of methodical procedures for structure determination. The focus is on precision and repeatability. The results are representative and can serve as a reliable basis for further decisions. It is the oldest of the three disciplines photogrammetry, SfM and SLAM, the latter two partly originating in it [16, 17].

4.4.2 Structure from Motion (SfM)

Structure from Motion, like photogrammetry, is used to estimate three-dimensional structures from two-dimensional image series. While photogrammetry is often performed in a known environment, sometimes even on a precise laboratory scale, the boundary conditions in the case of SfM are rarely controllable to the same extent. The SfM method also uses image series to create models of an object of interest. However, the sensor, usually a camera, is moving during the acquisition of the image series. While in photogrammetry often several fixed cameras from different viewing angles are used, in the SfM method the movement of the camera between two single images has to be considered when calculating the reconstruction. Even more, advanced SfM approaches can use images from cameras of different and at the same time unknown types, while the images can also be taken from completely different and sometimes unknown angles of view.

On the one hand, SfM can be seen as a sub-process of photogrammetry, being mostly focused on the determination of structure from images. However, in the scientific field of SfM numerous unique approaches and methods have been developed and refined. In contrast to photogrammetry, the focus is somewhat less on the absolute precision of the final result. Even though also SfM, like all disciplines, proceeds methodically to achieve the best possible outcome given the circumstances. SfM approaches allow for a much wider range of setups, the strength of SfM methods lies precisely in the inclusion of unknown measurement environments.

To determine the three-dimensional structure, SfM uses the motion parallax, the apparent displacement of an object in front of a background seen from different viewing locations. Here, the position of the object under investigation is not fixed, but is reconstructed from the image sequences. The displacement, rotation and tilt of the sensor is typically a known parameter. The information of the sensor movement is in some cases included as parameter to be refined while computing the three-dimensional structure from two-dimensional images.

The mathematical representation of humans perceiving the three-dimensional structure of an object with two eyes when seeing it can be formulated in an analogous way. The disciplines of stereo vision and computer vision form the artificial counterpart to the human depth perception. Here, the sensor is not shifted by a fixed amount, but the distance between two cameras or the human eyes is fixed. From a mathematical point of view, it makes no difference whether to use one camera in two positions or two or more cameras from different angles. With the restriction, that the observed scene is not allowed to change between capturing the images. Therefore, the respective research areas show strong overlaps. Although each discipline places the emphasis of the respective optimization strategies on different aspects, according to the corresponding research questions. Thus, the focus of the research shifts slightly between the disciplines. For projects to be implemented in practice and for specific tasks, it is therefore of high value to compare and systematically evaluate the approaches of all areas. It might be the most feasible way to build upon existing frameworks and combine several well-established methods from the different fields.

In summary, SfM focuses on the same object of investigation as photogrammetry, or can be seen as a part of it. However, the boundary conditions are less strict, SfM also works in more uncertain environments. The representation of the structure to be investigated can result in a somewhat thinner point cloud, the tolerances can be somewhat higher. In return, the user gets more freedom in using the methods and therefore needs less prior knowledge about the object to be investigated and the investigation environment [19].

4.4.3 Simultaneous Localization and Mapping (SLAM)

Simultaneous localization and mapping extend the problem of SfM in a certain way. While in SfM the sensor movement between two images is a usually a known quantity, in SLAM it is not specified. Thus, the problem is to reconstruct a three-dimensional structure from two-dimensional sensory information whose relative position to each other is also unknown. In other words, SLAM refers to the measuring of an environment, which can be described as its mapping, and the simultaneous acquisition of the sensor position, the localization of the sensor within the map [21, 22].

At first glance, this task poses a chicken-and-egg problem. If not only the position and viewing angle of a sensor in a given environment is not known, but also the entire geometry of this area is unknown, how can reliable statements be made? And exactly this question hints at the essential differences compared to the very exact photogrammetry as well as to the relatively exact SfM methods. In contrast to the other two areas, the statements, which are derived by SLAM methods about an object, are subject to much larger uncertainties.

However, SLAM opens up an extremely interesting area that is not easily accessible with other methods. SLAM does not usually examine a single object in a secured environment as photogrammetry does, nor does it examine a single object in a known space. Instead, as the naming part mapping also indicates, it is primarily concerned with the three-dimensional structural reconnaissance of larger contiguous areas. The approach in SLAM consequently uses the methods of photogrammetry and SfM and could possibly be classified as a subfield again. However, the object of investigation is of a fundamentally different nature and the methodological approaches used from the other fields have at least been transferred to be well adjusted to problems occurring in SLAM. And key components to solve the core issues have been developed within this field of research, therefore SLAM, like the aforementioned disciplines, is rightly to be seen as an independent domain.

The problem of assessing the geometry of an entire environment while simultaneously determining the position, rotation and inclination of the sensor relative to this environment can actually be solved in the theoretical case. And also, implementations in some special environments show very good results. However, new and unknown environments are still challenging to solve and are subject to a lot of ongoing research.

As in the other methods, the acquired images are standardized. Subsequently, a correlation between the individual images can be established by recognizing similar features. Due to this technique it follows directly that the displacement of the sensor must be within limits between two pictures to allow for large overlaps in the depicted structure. Many unknown parameters are determined and optimized simultaneously in the processes of determining the structure from the images. Even with modern computer systems this is a capacity problem. The result is often a compromise between accuracy and computing time. The ideal implementation remains a subject of current research and can be a decisive competitive advantage in commercially oriented applications.

Since, in contrast to photogrammetry and SfM, the boundary conditions are not clearly defined, SLAM corresponds to a real-time method. Meaning that the boundary conditions are only determined during the image evaluation. The obtained results typically end in thinner point clouds of the three-dimensional objects under investigation. As the examined scenes can be very large, the absolute number of determined points can nevertheless be much higher. Typically, SLAM algorithms operate in a multistage manner, an initially identified sparse point cloud can subsequently be transformed to being denser. After the connection between the individual images is well established, details can be extracted in a second stage. Nevertheless, this is not to be confused with the exact photogrammetry. Thus, the final results of SLAM might be orders of magnitude less precise due to the many unknowns. However, SLAM algorithms achieve impressive results in modern implementations [23–25].

Finally, if we look at the different fields of photogrammetry, SfM and SLAM, it becomes clear why they are so close to each other. All of them are suitable to infer a three-dimensional structure on the basis of two-dimensional data. However, the differences emerge when defining the object or environment in more detail. Photogrammetry stands for reliable, repeatable, precise results. SfM also works in unfamiliar environments using a moving sensor while knowing its relative position. Finally, in SLAM methods, the position of the sensor is initially unknown and is included as parameter within the toolchain. The free parameters and uncertainties increase, but also the possible use cases expand. Methods from all fields have successfully been implemented in commercial applications. In the end, the selection will have to be based on the problem to be investigated and it might include several partial aspects from all areas.

4.5 Image-Based 3D Reconstruction

The following chapter describes image-based 3D reconstruction. In this procedure, camera poses are calculated by finding 2D point correspondences between the individual images relative to each other. The camera poses form the basis for the generation of the 3D point cloud [26]. Figure 4.3 shows the individual steps of the 3D reconstruction.

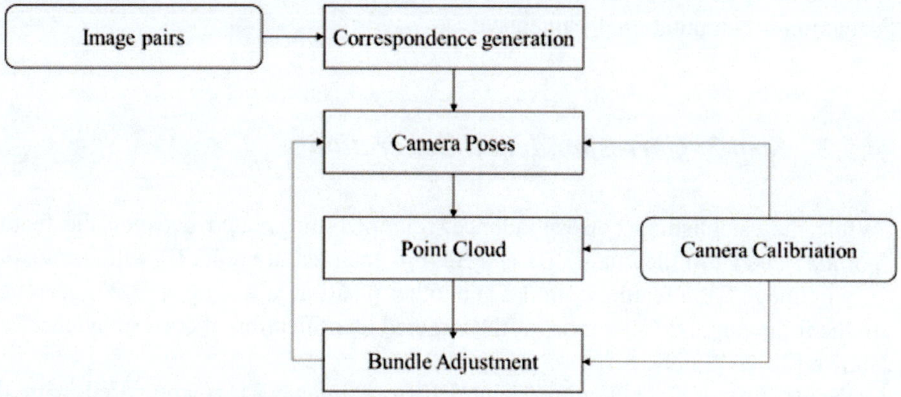

Fig. 4.3 Steps on the example for optimization on a pair of images

4.5.1 Step 1: Feature Extraction

The detection of corresponding points in different images is an elementary part of 3D reconstruction, object recognition, navigation of robots and many more. The method uses feature detectors or feature descriptors to find and describe salient points. The task of the feature detector is to find as many distinctive points (feature points) as possible [27]. A feature point is a point with the following properties:

- This feature point can be described by feature descriptors.
- The point can be found in images recorded in slightly different shooting conditions.
- A point that is described as accurately as possible by its environment and is distinguishable from other points.

Feature descriptors are high-dimensional vectors used to store information about the environment of feature points. Known algorithms for computing features are named below:

SIFT: (scale invariant feature transform) The algorithm provides a feature detector and descriptor. The detector and descriptor are scale and rotation invariant and limitedly robust against different illumination situations.

SURF: (Speeded up robust features): The algorithm also has a feature detector and descriptor. The detector is faster than the SIFT detector. The algorithm has largely the same advantages and disadvantages as the SIFT algorithm.

KAZE: The algorithm follows the same approaches as SIFT and SURF, but a nonlinear filter is used for the detector, being better at detecting edges and corners. For the descriptor, a modification of SURF is used. Due to the nonlinear filter, the detection is computationally intensive.

4.5.2 Step 2: Correspondence Generation

While the establishing correspondence, relations are created between the feature point F1 and F2 of the images B1 and B2. For each feature point F1 with its position x, y in image B1, a feature point F2 at another position x, y in image B2 is searched. In the following, two solutions for the targeted identification of correspondences are shown [28].

Brute Force: Each feature point F1 from image B1 is compared with the feature points from image B2. In this algorithm, the computational effort increases quadratically with the number of features to be compared.

FLANN (Fast Library for Approximate Nearest Neighbors) is an algorithm for large datasets with high dimensional features. The nearest neighbor search algorithm is mainly used in the field of image recognition, compression and pattern recognition. There are different approaches to implement this algorithm. The software library FLANN uses k-d or k-means trees to store the individual feature points.

Each of the images to be compared has its own tree. Searching in trees for the corresponding points is much more efficient than the Brute Force approach. In this case, the computational effort increases logarithmically with the number of feature points to be compared. Once a possible corresponding feature point is found, it has to be decided whether the correspondence between the feature point from image B1 and the other feature point from image B2 is of the necessary quality.

For images with a lot of structure it can happen that a wrong correspondence between feature point F1 from image B1 to feature point F2 from image B2 is found. Also, for one feature point several other feature points might be found in the other image. In this case the distance between the individual feature points is used as a criterion to asses quality.

4.5.3 Step 3: 3D Reconstruction

In this chapter the information from the discovery of feature points and mapping is used to calculate the 3D point cloud.

Camera Model (Inner Camera Parameters)
For a true-to-scale 3D reconstruction, a model must be defined that describes how a camera captures an image. That is, how a point in the real world is mapped onto the sensor of a camera. For cameras (smartphone, reflex cameras and alike) which produce 2D images, typically a simple pinhole model is sufficient. The pinhole model describes the mathematical relationship between the coordinates of a point in three-dimensional space and the mapping to the image plane inside a pinhole camera without any side effects [29]. Figure 4.4 illustrates the pinhole model.

The figure shows the projection of the point p_r on p_s. The z-axis corresponds to the optical axis with the main point p_H over the camera center c. The distance between c and p_H is the focal length.

The camera aperture is representing by a point as a simplification of any lenses focusing the light. Thus, no geometric distortions or blurring of objects caused by lenses or finite aperture sizes are considered in the model. The pinhole camera model

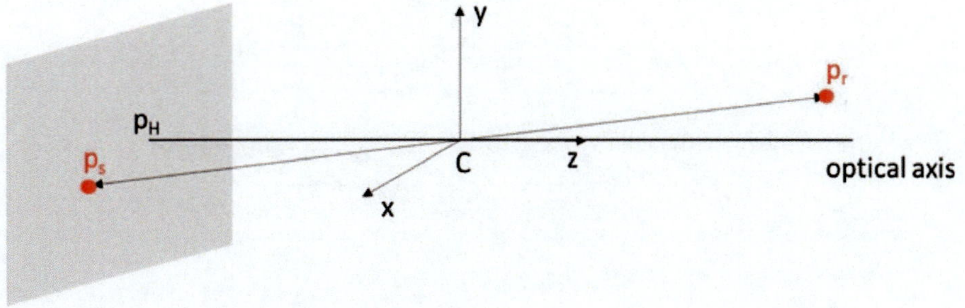

Fig. 4.4 Projection between real point p_r and sensor point p_s

can be seen as an approximation of the mapping from a 3D scene to a 2D image. Its validity depends on the quality of the camera being is used. The higher the distortion in the outer areas of the image caused by the lenses of the camera, the higher the deviation between camera model and image. Unfortunately, there is no ideal camera, each camera shows different deviations over the range of an image.

To account for the deviations from reality, a calibration process is used. During the calibration of a single pinhole camera, intrinsic parameters are determined, describing largely the path of the light on the inside of the camera. For this purpose, a checkerboard is recorded from different angles of view. The intersection points of the individual squares are determined via feature recognition. As the checker board consists of straight lines, the individual points must also result in pattern of straight lines on the image. If a point deviates from this pattern of straight lines, the difference is identified as a distortion. As the distance from the camera to the checkerboard is known as well as the dimensions of the board itself, the relation between centimeters and pixels can also be calculated.

Outer Camera Parameters

The outer camera parameter describes the difference between two camera positions. The delta is expressed in two movements, the translation (displacement) and rotation on the corresponding axes. With the outer camera parameters, it is possible to convert from one camera position into the another.

Epipolar Geometry and the Fundamental Matrix

Epipolar geometry as sub-field of geometry forms the mathematical framework to describe the geometric relationships between two different camera images as well as the relationships of individual objects within these pictures [30]. The Fig. 4.5 shows the epipolar geometry.

With the camera parameters the center space c_1/c_2 and the and the mapping from m_1/m_2 of the image plane S_1/S_2 to c_1/c_2 is known. The 3D point M is projected over image point m_1/m_2 onto the two centers of the cameras over the epipolar plane T. The straight line from C_1 to C_2 is called the baseline. The intersections of the baseline

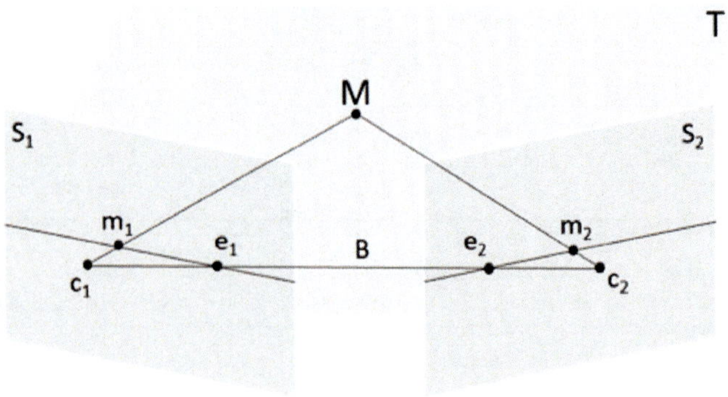

Fig. 4.5 Epiploar geometry

with the image plane S_1/S_2 are called epipoles e_1/e_2. The epipolar lines l1/l2 is the connection between the epipoles e_1/e_2 and the image point m_1, m_2. The epipoles is the projection into the respective other image planes.

Properties of the Essential and Fundamental Matrices

Both matrices describe the relation between corresponding points in two images. The fundamental matrix (F) describes the points in pixel coordinates, the essential matrix (E) describes the points in a normalized image coordinate system. For deriving the normalized image coordinates, the origin is placed the optical center of the image.

The two matrices are related to each other as follows: $E = K' * F * K$. The matrix K and K′ are the intrinsic matrixes of the two cameras, which is generated by establishing its calibration. The fundamental matrix has seven degrees of freedom and the essential matrix five degrees of freedom, since the camera parameters are considered here. If the camera poses of the two cameras are not known, the fundamental matrix can be calculated by using eight points showing in both of the images. If the camera matrices are known, five points are sufficient.

The Camera Pose

In the fundamental matrix the relative pose of two cameras is described, not accounting for any scaling factors from one to the other. In a first step the fundamental matrix between two pictures can be derived using a maximum of eight shared points as explained. If this fundamental matrix is multiplied with the calibration data of the individual cameras, four essential matrices are created, whereby only one of these gives a valid representation of the image. The correct essential matrix is derived by testing with other point correspondences. The final matrix describes the relative camera pose between two cameras.

Through this procedure, however, the scaling factor is not yet known. For the initial start of a 3D reconstruction, a camera position is set as the origin of the world coordinate system. If further cameras are added to the 3D model, the scaling between the individual images must be adjusted. The problem is described as the perspective-n-point problem. With the solution of the perspective-n-point problem, the absolute poses of the cameras can be calculated. For solving it, the images and 3D world points as well as the calibration matrix of each camera are necessary. Using these inputs, the position and orientation of the camera can be determined including the scaling. The algorithm also compensates for inaccuracies in the former calculation of the relative camera poses. The algorithm must thereby find a solution for world points, camera poses, camera calibration to the point correspondences within the data [31].

Bundle Adjustment

In any reconstruction method that uses points or a camera poses with more than two images to calculate a 3D model, small errors occur when joining the individual images. The errors have different reasons, like inaccurate localization of the feature points, inaccurate camera calibration, wrong point correspondences or others. With each addition of more images and their relative poses, all these relatively small errors add up effecting the final result significantly. Bundle adjustment tries to find the same feature points not only in one image pair, but in several images. The discovered additional feature points are then used for the bundle adjustment. As a start the algorithm establishes an initial 3D model. This model is now optimized including all distinctive feature points and all camera poses [32].

Combing bundle adjustment and a cost function, the mathematic camera model is further optimized in respect to the projection of the real-world points and their representation in the camera sensor. With this approach, a back projection error can be calculated and thus an indication of the accuracy of the mapping of the world points to the feature points in the images can be obtained. Depending on the input parameters, different parameters of the model are optimized in the cost function. If a calibrated camera is used, the intrinsic parameters are not adjusted, in this case the algorithm optimizes camera poses and world points. Outliers in the images cause world points to be outside the valid range in the 3D model. These outliers create a high back projection error and thus a large cost in the algorithm. In turn these large costs prevent the algorithm from finding a correct optimum. The problem can be addressed by introducing a thresholding cutting large costs due to outliers.

Bundle adjustment reduces the error significantly, but small errors due to the summation of the individual poses over several images remain. The deviation becomes apparent when the image series depicts its starting points again and the representation of the same points in the real world do not match in the model. Strong discrepancies indicate a particularly high back-projection error.

With an explicit loop closure, the problem of drifting apart over larger models of several hundred images is solved. For this purpose, in a first step the poses at both ends showing the same points in the real world are slightly modified causing them to fit together. In the next step, the individual poses are optimized a second time using a specifically adjusted cost function.

The more often a loop closure occurs in the image series, meaning the camera is crossing a previously shown path again, the faster and more accurately the deviation of the individual camera poses can be corrected. Another benefit is also the significantly reduced computation time, but it has to be weighed against the increase in measurement time [33].

4.6 3D Reconstruction in Production Environments

The following chapters describe the testing of scanning a factory with different configurations of photogrammetry. The section "Definition of the test environment" describes the main test cases used to test each configuration.

4.6.1 Definition of the Test Environment

After establishing the basics of reconstructing three-dimensional environments using photogrammetry, its real-world applications can be examined by defining several test cases.

Scenario 1: Detection of depths above ten meters

Production halls form large indoor environments, depths of up to and more than ten meters are often found.

Scenario 2: The door passageway

In contrast to the halls, passageways describe a narrow transition from one production hall to another or form the passage into a staircase.

Scenario 3: The homogeneous structure

This case covers the often homogeneous surfaces of machines, floors and walls in production being difficult to handle for some algorithms.

Scenario 4: The long corridor

A long aisle simulates the dimensions of production facilities. It is necessary to assess the deviation between start and end of measurement, by measuring a long distance the displacement of the same points is displayed. Additionally, the measured environment consists of mainly white walls, causing further difficulty to identify unique feature points.

General setup of the camera

To compare between individual technologies, the following parameters are defined for all tests:

- ISO value: 100
- White balance: no automatic white balance
- Compression: low to no compression
- Light source: no additional artificial light source.

4.6.2 Monocular Method

The method uses a single camera to capture images. Either the images are extracted from a video stream or individual images are recorded [29–31].

Scenario 1: Detection of depths above ten meters

The depth of ten meters is very well represented when capturing a large number of images containing an overlap of at least 60%.

Scenario 2: The door passageway

Reconstructing the three-dimensional structure in this narrow case scenario causes a problem in the test case. While the hall at the beginning of the recording is represented very well, the camera pose is lost when crossing the door. The following picture shows a coffee kitchen being completely wrong integrated into the 3D coordinate system. A new fragment is placed on the left behind the wall instead of framing the scene.

Scenario 3: The homogeneous structure

Homogeneous structures are only reconstructed well if the images are taken in very close proximity. If the images are taken at a normal distance, considerable noise is visible in the point cloud.

Scenario 4: The long corridor

For a long and white corridor using the monocular method, the reconstruction results at best at an accuracy of five centimeters at ten meters distance.

Summary

3D reconstruction with a single camera in a production environment shows considerable limits. It might be applicable for the scan of workstations of two to three meters. There must be a large number of distinctive points in the individual images in order to correctly determine the geometric properties while processing the images. When using a smartphone, the dimensions of the scan are reduced even further. Here, a maximum of only two meters is recommended. With the pinhole model using a camera with a small opening angle is a major disadvantage, further advancements might be possible with wide angle cameras [37–39].

4.6.3 Stereo Images (Stereo Camera)

Recording with a stereo camera allows the calculation of depth maps based on two images taken from different angles at the same time. The distance between cameras, the baseline is key in this setup. It is calculated using the intrinsic and extrinsic calibration. With larger baselines greater depths can be better resolved, but in turn closer elements are not resolved as well. When using a stereo camera with a baseline of ten centimeters depths up to five meters are well represented in reconstruction models. The setup is therefore well suited within close ranges like depicting office spaces. However, depths greater than five meters cannot be accurately calculated with the camera and the setup has to be adjusted. For the measurement of production facilities, a camera with a baseline of 60 cm leads to good representations of depths of ten meters and more.

Before calculating the camera pose, firstly the two images have to be converted to a single image showing the scene and a depth image. It is common practice to calculate the depth image for the left perspective. The great advantage of stereo cameras is to yield information about the scene and the depth at the same time. Reconstructing the camera pose and the 3D model is therefore shortened and is likely to lead to better results [40–42].

Scenario 1: Detection of depths above ten meters

The correct calculation of a depth of ten meters can be well accomplished with the setup as described. The accuracy is in the range of one to two centimeters.

Scenario 2: The door passageway

The door passage is well reconstructed by the system. This might be especially helpful in some scenarios as it provides a likely point to perform loop closures.

Scenario 3: The homogeneous structure

Homogeneous structures are reconstructed mostly well standing directly before them and also from a greater distance. But a major problem emerges. If the camera passes close to a machine with a homogeneous surface as is likely to happen when scanning within factories, too few feature points are recognized. Without sufficient feature points the camera pose is lost and a further synchronization is no longer possible.

Scenario 4: The long corridor

For the white long corridor, the accuracy of the reconstruction does not surpass two to three centimeters at ten meters distance. Compared to the monocular method, the improvement of the accuracy originates in obtaining the depth images in the first step (Fig. 4.6).

Summary

The calculation of the three-dimensional model is improved, but the inaccuracies are still too high for a general recommendation considering the use case of reconstructing a production environment. Further optimization measures like using two

Fig. 4.6 Depth image from stereocamea in front of a machine

independent algorithms for the calculation of the camera pose can improve accuracy to some degree, yet not sufficiently. Using cameras with a wider angle of opening also increases the quality of the reconstruction. However, in total the system does not meet the requirements set [43–45].

4.6.4 Spherical Images

A spherical image is a panoramic image that includes all angles of view in every single shot. The opening angle is therefore 360° in width and 180° in height. With this information the complete surface of a sphere is covered. The point of capture is in the center of the sphere.

A rectangular projection can be used to store the two-dimensional representation of spherical images. This projection maps the horizontal 360- and 180-degree view in a two to one aspect ratio. The zenith and nadir are drawn over the complete upper and lower image edges, respectively. The parts in the center of the image which are close to middle horizon are displayed largely without distortion [46–48].

These images have an angle of aperture of 360° and thus result in many more feature points which can be tracked from one image to the next. Due to the linear relationship of the pixel size to the focal length and distance, no complex calibration of the camera is necessary.

Scenario 1: Detection of depths above ten meters

Due to the large opening angle, many feature points are detected, resulting in a good calculation of the depth. The accuracy is about one to two centimeters at ten meters distance.

Scenario 2: The door passageway

The large opening angle allows a problem-free calculation of the camera pose at the doorway because feature points from the old and the new room can be used at the same time.

Scenario 3: The homogeneous structure

With spherical cameras measuring only homogeneous structure still poses a problem. However, if cameras with a high resolution (width greter 6000 pixel) are used, minimal sub-structures can be resolved in the close-up scenario and the number of feature points and quality of the reconstruction increases.

Scenario 4: The long corridor

Because of the large aperture angle long corridors can be reconstructed without any problems. Constantly enough feature points in the images are identified within the image series. Resolutions of one centimeter at a depth of ten meters can be achieved.

Fig. 4.7 Shows equiractangular representation of tea kitchen

Summary

With a spherical camera the 3D model was reconstructed very well. Due to the large opening angle of 360°, many feature points are detected in all directions. The points allow for a very good bundle adjustment, because there are always more common points than the minimum requirement of eight (Fig. 4.7). Loop closure leads to an even better refined result [49–54].

4.7 Conclusions and Outlook

At the beginning, requirements were defined for the recording system, two important ones allowing for fast recording and a simple handling. Due to this, even though achieving high accuracy, a laser scanner is often not ideal in the case of measuring production environments. Photogrammetry on the other hand is well suited to meet these requirements, but has to prove its applicability and accuracy [55–57].

Four test scenarios were created to test the performance of the system in typical field situations. The monocular scan method works very well up to a maximum of two to three meters. Minor improvements can be achieved by adding a depth sensor as often found in newer smartphones. With a stereo camera and a baseline of 60 cm, depth images can be directly calculated for every shot, leading to further optimization of the 3D model. However, the small opening angle causes problems, if the stereo camera is moved directly before a machine. The faster a scan is performed while being in front of the machine, the greater the probability that the camera pose will be lost. In this case the reconstruction cannot be continued correctly [58].

In the last step, spherical images were used. The major advantage here is the large opening angle of 360 degrees. If the wide opening angle is combined with a high resolution, many feature points are found. Moving close to the front of the machines does not cause any problems, feature points to assess the correct camera pose are simply used from the other side of the field of view. Due to the large number of features, a very good bundle adjustment is possible. Further improvements in the performance can be accomplished by an explicit loop closure, but this is not necessary in all cases [59].

In summary, the requirements could be realized very well using a spherical camera. Losses of the camera pose are rare. Handling the camera is easy and three-dimensional reconstruction from the images is robust. Thereby the execution of the measurement is also suitable for untrained personal. The commercial spherical cameras achieve higher and higher resolutions and already meet the demand for sufficient resolution of homogeneous structures.

A higher resolution often also leads to higher levels of compression when storing images in videos. However, too much compression of the images has a negative consequence on the detection of features in the images. This results in higher levels of noise which become visible quickly on white walls or other homogenous surfaces. In the reconstruction these are not a straight but sinusoidal wave become visible.

An important factor in complete 3D reconstructions based on images is a very good recording of the entire environment in constant conditions. The images must be taken without automatic white balance and without autofocus. A small level of compression of the images is desirable, it is best to process the images in raw format. Anything not being present during the acquisition of the data cannot easily be reconstructed due to the many steps involved and attempting it before the final result is created might significantly degrade the quality of the 3D model.

With a spherical camera and a suitable image processing pipeline, a 3D model can be created fast and efficiently by simple means, if the accuracy of the model does not have to be in the range of millimeters. The model is apt to serve as basis for further analysis such as building a CAD model and subsequent simulations are possible at any time.

A further step would be the reconstruction of gaps within the 3D model at a later stage. Gaps originate largely from homogeneous surfaces. One approach would be to create a surface based on the existing points and extend it to cover any missing features. As edges are very well represented within the 3D model this approach is expected to be straight forward and to lead to good results [60].

References

1. Iqbal M, Hashmi MSJ (2001) Design and analysis of a virtual factory layout. J Mater Process Technol 118:403–410
2. 3D reconstruction. https://en.wikipedia.org/wiki/3D_reconstruction
3. Ein Rückblick auf das 3D-Laserscanning: von den 50er Jahren bis heute. https://mep.trimble.com/de/resources/blogs/die-evolution-des-3d-laserscannings-2
4. Laser scanning Terrestrisches Laserscanning. https://en.wikipedia.org/wiki/Laser_scanning
5. Photogrammetry. https://en.wikipedia.org/wiki/Photogrammetry

6. 3D Scanning. https://en.wikipedia.org/wiki/3D_scanning
7. Besbes M, Zolghadri M, Afonso RC, Masmoudi F, Haddar M (2020) 3D facility layout problem. J Intell Manuf. https://doi.org/10.1007/s10845-020-01603-z
8. Patraucean V, Armeni I, Nahangi M, Yeung J, Brilakis I, Haas C (2015) State of research in automatic as-built modelling. Adv Eng Inform 29:162–171
9. Carvalho LE, von Wangenheim A (2019) 3D object recognition and classification: a systematic literature review . Pattern Anal Appl 22:1243–1292. https://doi.org/10.1007/s10044-019-00804-4
10. Javaid M, Haleem A, Kumar L (2018) Current status and applications of 3D scanning in dentistry. Clinical Epidemiology and Global Health, Elsevier, a division of RELX India, Pvt. Ltd on behalf of INDIACLEN, 6 July 2018
11. Moona D, Chunga S, Kwonb S, Seoc J, Shin J (2019) Comparison and utilization of point cloud generated from photogrammetry and laser scanning: 3D world model for smart heavy equipment planning. Autom Construct 98:322–331
12. Li F, Kim M-K (2020) Mirror-aided registration-free geometric quality inspection of planar-type prefabricated elements using terrestrial laser scanning. https://doi.org/10.1016/j.autcon.2020.103442
13. Lindskog E, Berglund J, Vallhagen J, Johansson B (2016) Layout planning and geometry analysis using 3D laser scanning in production system redesign. In: 6th CIRP Conference on Assembly Technologies and Systems (CATS)
14. Bubaker-Isheil H, Hennebelle F (2019) Jean-Francois Fontaine, simple large scale 3D scanner. In: 13th CIRP conference on intelligent computation in manufacturing engineering. CIRP ICME '19
15. Rebolj D, Pučko Z, Babič NČ, Bizjak M, Mongus D (2017) Point cloud quality requirements for Scan-vs-BIM based automated construction progress monitoring. Autom Construct 84:323–334
16. Shashi M, Jain K (2007) Use of photogrammetry in 3D modeling and visualization of buildings. ARPN J Eng Appl Sci
17. Mancini F, Salvini R (2019) Applications of photogrammetry for environmental research. ISPRS Int J Geo-Inf 8:542
18. Emmanuel PB (1999) A comparison between photogrammetry and laser scanning. ISPRS J Photogrammetry Remote Sens 54(2–3):83–94
19. Hahner M, Bountouris P, Varesis O (2017) Simulating structure-from-motion. arXiv:1710.01052v1 [cs.CV] 3 Oct 2017
20. Schnitzer F, Sonnenburg A, Janschek K, Willich G (2012) Bildbasierte Slam-Relativnavigation und 3drekonstruktion für das on-orbit-servicing von unbekannten und unkooperativen raumflugkörpern. Deutscher Luft- und Raumfahrtkongress
21. Kuo B-W, Chang H-H, Chen Y-C, Huang S-Y (2011) A light-and-fast SLAM algorithm for robots in indoor environments using line segment map. https://www.researchgate.net/publication/228921788. J Robot
22. Prasad V, Das D, Bhowmick B (2019) Epipolar geometry based learning of multi-view depth and ego-motion from monocular sequences. arXiv:1812.11922v3 [cs.RO] 7 Jan 2019
23. Baumann S (2011) Modulare Sensorfusion zur dynamischen Anreicherung einer SLAM-Karte fur die Verwendung auf mobilen Robotern, Universität Bielefeld, Technische Fakultät, 20. Juni 2011
24. Price E (2013) Evaluierung von Verfahren zum optischen Lokalisieren und Kartographieren (SLAM) mit Eignung für den Einsatz auf UAVs, Universität Stuttgart, Institut für Parallele und Verteilte Systeme, 17. Januar 2013
25. Lemaire T, Lacroix S (February 2007) SLAM with panoramic vision. J Field Robot 24(1–2):91–111
26. Ham H, Wesley J, Hendra H (2019) Computer vision based 3D reconstruction: a review. Int J Electr Comput Eng 9(4):2394. https://doi.org/10.11591/ijece.v9i4.pp2394-2402

27. Kumar G, Bhatia PK (2014) A detailed review of feature extraction in image processing systems. In: IEEE fourth international conference on advanced computing & communication technologies. At: Rohtak, Haryana, India, February 2014. https://doi.org/10.1109/ACCT.2014.74
28. Dong Y, Fan D, Ma Q, Ji S (2020) Image feature matching via parallax mapping for close range photogrammetric application. IEEE Access (99):1–1. https://doi.org/10.1109/ACCESS.2020.2973723
29. Percoco G, Guerra MG, Sánchez-Salmerón A-J, Galantucc LM, (2017) Experimental investigation on camera calibration for 3D photogrammetric scanning of micro-features for micrometric resolution. Int J Adv Manuf Technol 91(9). https://doi.org/10.1007/s00170-016-9949-6
30. Basta T (June 2013) Does the fundamental matrix define a one-to-one relation between the corresponding image points of a scene? Int J Image Graph 1(3):125–128. https://doi.org/10.12720/joig
31. Melekhov I, Ylioinas J, Kannala J, Rahtu E (2017) Relative camera pose estimation using convolutional neural networks. In: International conference on advanced concepts for intelligent vision systems. https://doi.org/10.1007/978-3-319-70353-4_57
32. Hansch R, Drude I, Hellwich O (2016) Modern methods of bundle adjustment on the GPU. In: ISPRS annals of the photogrammetry, remote sensing and spatial information sciences, volume III-3, 2016 XXIII ISPRS Congress, 12–19 July 2016, Prague, Czech Republic
33. Cremers D (2019) Chapter 7 bundle adjustment & nonlinear optimization multiple view geometry. Technical University of Munich Summer 2019. https://vision.in.tum.de/_media/teaching/ss2019/mvg2019/material/multiviewgeometry7.pdf
34. Yu T, Zou J-H, Song Q-B (2017) 3D reconstruction from a single still image based on monocular vision of an uncalibrated camera. ITM Web of Conferences 12, 01018
35. Eckhof M (2015) Untersuchung der Eignung einer Handykamera fur die dreidimensionale ¨ Objektrekonstruktion. Leibniz Universität Hannover, Institut fur Photogrammetrie und Geoinformationen
36. Atapour-Abarghouei A, Breckon TP (2018) Real-time monocular depth estimation using synthetic data with domain adaptation via image style transfer. In: 2018 IEEE/CVF conference on computer vision and pattern recognition
37. Dygoduk S (2012) Entwicklung einer Android App für 3D Rekonstruktion, Hochschule für Angewandte Wissenschaften Hamburg, Fakultät Technik und Informatik, 30 August 2012
38. Manthe S (2014) 3D-Rekonstruktion aus monokularen Bilderserien, Universität Koblenz, September 2014
39. Pollefeys M, Nist´er D, Frahm J-M, Akbarzadeh A, Mordohai P, Clipp B, Engels C, Gallup D, Kim S-J, Merrell P, Salmi C, Sinha S, Talton B, Wang L, Yang Q, Stewénius H, Yang R, Welch G, Towles H (2008) Detailed real-time urban 3D reconstruction from video. Int J Comput Vision
40. Blank C (2012) Generierung von Tiefenbildern mittels Stereoskopie, Hochschule für Angewandte Wissenschaften Hamburg, Fakultät Technik und Informatik, 30 August 2012
41. Schönberger JL, Zheng E, Pollefeys M, Frahm J-M (2016) Pixelwise view selection for unstructured multi-view stereo. https://www.researchgate.net/publication/305655847
42. Chang A, Dai A, Funkhouser T, Halber M, Nießner M, Savva1 M, Song S, Zeng A, Zhang Y (2017) Matterport3D: learning from RGB-D data in indoor environments. arXiv:1709.06158v1 [cs.CV] 18 Sep 2017
43. Breuer T, Bodensteiner C, Arens M (2014) Low-cost commodity depth sensor comparison and accuracy analysis. In: Proceedings of SPIE the international society for optical engineering. https://doi.org/10.1117/12.2067155
44. Maier R (2019) High-quality 3D reconstruction from low-cost RGB-D sensors. Technische Universität München, Department of Informatics
45. Zhang L, Chen D, Liu W (2016) Fast plane segmentation with line primitives for RGB-D sensor. Int J Adv Robot Syst 1–8
46. Ki0m H, Hilton A (201) 3D modelling of static environments using multiple spherical stereo, centre for vision, speech and signal processing. University of Surrey Guildford, GU2 7XH, Surrey, UK (2010). https://www.researchgate.net/publication/228593686

47. da Silveira TLT, Jung CR (2019) Dense 3D scene reconstruction from multiple spherical images for 3-DoF+ VR applications. In: IEEE conference on virtual reality and 3D user interfaces, Oska
48. Xu J, Stenger B, Kerola T, Tung T (2016) Pano2CAD: room layout from a single panorama image. arXiv:1609.09270v2 [cs.CV] 30 Sep 2016
49. Abate D, Toschi I, Sturdy-Colls C, Remondino F (2017) A low-cost panoramic camera for the 3D documentation of contaminated crime scenes. In: The international archives of the photogrammetry, remote sensing and spatial information sciences, volume XLII-2/W8, 2017 5th international workshop lowcost 3D—sensors, algorithms, applications, 28–29 Nov 2017, Hamburg, Germany
50. Bruno N, Roncella R (2019) Accuracy assessment of 3D models generated from google street view imagery. In: The international archives of the photogrammetry, remote sensing and spatial information sciences, volume XLII-2/W9, 2019 8th international workshop 3D-ARCH "3D virtual reconstruction and visualization of complex architectures, 6–8 Feb 2019, Bergamo, Italy
51. Perfetti L, Polari C, Fassi F (2017) Fisheye photogrammetry: tests and methodologies for the survey of narrow spaces. In: The international archives of the photogrammetry, remote sensing and spatial information sciences, volume XLII-2/W3, 2017 3D virtual reconstruction and visualization of complex architectures, 1–3 March 2017, Nafplio, Greece
52. Perfetti L, Fassi F, Rossi C (2019) Fisheye Photogrammetry to generate low-cost DTMS. In: The international archives of the photogrammetry, remote sensing and spatial information sciences, volume XLII-2/W17, 2019 6th international workshop lowcost 3D—sensors, algorithms, applications, 2–3 Dec 2019, Strasbourg, France
53. Caracotte J, Morbidi F, Mouaddib E (2021) Photometric stereo with Twin-Fisheye Cameras. In: 25th international conference on pattern recognition, Jan 2021, Milan, Italy, pp 5270–5277
54. Chiappini S, Fini A, Malinverni ES, Frontoni E, Racioppi G, Pierdicca R (2020) Cost effective spherical photogrammetry: a novel framework for the smart management of complex urban environments. In: The international archives of the photogrammetry, remote sensing and spatial information sciences, volume XLIII-B4–2020, 2020 XXIV ISPRS Congress (2020 edition)
55. Han B, Paulson C, Dapeng Wu (2011) 3D dense reconstruction from 2D video sequence via 3D geometric segmentation. J Vis Commun Image Represent 22(5):421–431
56. Runceanu LS, Becker S, Haala N, Fritsch D (2017) Indoor point cloud segmentation for automatic object interpretation, 37. Wissenschaftlich-Technische Jahrestagung der DGPF in Würzburg – Publikationen der DGPF, Band 26, 2017
57. Zhang Y, Bai M, Kohli P, Izadi S, Xiao J (2017) DeepContext: context-encoding neural pathways for 3D holistic scene understanding. arXiv:1603.04922v4 [cs.CV] 16 Aug 2017
58. Sedlacek M (2012) Sparse approximate inverses for preconditioning, smoothing, and regularization. TECHNISCHE UNIVERSITÄT MÜNCHEN Lehrstuhl für Informatik mit Schwerpunkt Wissenschaftliches Rechnen
59. Al-Nuaimi A (2016) Methods of point cloud alignment with applications to 3D indoor mapping and localization. Technische Universität München, Fakultät für Elektrotechnik und Informationstechnik Lehrstuhl für Medientechnik
60. Zhang X, Zhao P, Hu Q, Wang H, Ai M, Li J (2019) A 3D reconstruction pipeline of urban drainage pipes based on multiview image matching using low-cost panoramic video cameras. Water, 9 Oct 2019

Chapter 5
Machine Learning in Manufacturing in the Era of Industry 4.0

Markus Sommer and Josip Stjepandić

5.1 Introduction

Artificial Intelligence (AI), defined as a system's ability to correctly interpret external data, to learn from such data and to use those learnings to achieve specific goals and tasks through flexible adaptation, has attracted immense attention among academics, business managers, entrepreneurs, and politicians alike [1]. AI is the study of features of human activities, constructing a certain intelligent system, to make computers complete the tasks that only human is able to do in the past, and to apply computer hardware and software to simulate the underlying, theories, approaches and techniques of human behavior [2].

Machine learning (ML) as a subset of AI is an overwhelming trend, both in research and industrial applications. Although many of the machine learning ideas have been around for many years, the recent breakthroughs are based on several advances. One is the availability of large datasets with labeled data. Another is the availability of fast specialized processors such as graphics processing units (GPUs). Progress is fueled by a deeper understanding of building models and learning from data, as well as some new techniques that have brought everything together [3].

Machines themselves identify necessary knowledge from the users, data, and environments surrounding them [4]. The learning process can be supervised (in the presence of data and controlled learning mechanism), unsupervised or hybrid [5]. Applying statistical methods to structured and unstructured databases allows to

M. Sommer (✉)
isb innovative software businesses GmbH, Otto-Lilienthal-Strasse 2, 88046 Friedrichshafen, Germany
e-mail: sommer@isb-fn.de

J. Stjepandić
PROSTEP AG, Dolivostr. 11, 64293 Darmstadt, Germany
e-mail: josip.stjepandic@prostep.com

J. Stjepandić et al. (eds.), *DigiTwin: An Approach for Production Process Optimization in a Built Environment*, Springer Series in Advanced Manufacturing,
https://doi.org/10.1007/978-3-030-77539-1_5

extract previously unknown patterns and laws to generate new knowledge. This enables the formation of prediction models for data-based and computer-aided prediction of future events [6]. Deep learning (DL) has solved problems beyond the realm of traditional, hand-crafted machine learning algorithms and captured the imagination of practitioners who are inundated with all types of data. Further evolution facilitates some new automatic architecture optimization protocols that use multi-agent approaches [7]. Focusing the view on manufacturing, results of a recent study suggest that 75 percent of the possible research domains in ML-aided production planning and control (ML-PPC) are barely explored or not addressed at all [8].

The productive use of ML requires the expertise of a data scientist and thorough knowledge of the manufacturing processes. Small and medium-sized companies that specialize in certain high value-added, variant rich production processes often lack an in-house data scientist and therefore miss out on generating a deeper data-driven insight from their production data streams. Recent research proposed a three-step machine learning methodology to empower process experts with limited knowledge in machine learning: (1) data exploration through clustering, (2) representation of the production systems behavior through specially structured neural networks (NN) and (3) querying this representation through evolutionary algorithms to achieve decision support through online optimization or scenario simulation [9].

A review on ML methods for manufacturing is given by Fahle et al. [10]. The survey was conducted with respect to criteria of subtopic, specific application and algorithm (Table 5.1). NN are models consisting of nodes, weights and layers and are intended to mimic a learning approach similar to the brain. Whereas tree algorithms are architectures with branches that inherit information about an item, up to its leaves which contain the target of the regarded item. Besides models from supervised learning, there are techniques from unsupervised learning such as singular value decomposition (SVD) and principal component analysis (PCA) that can be used to reduce the dimension of the data. A frequently used algorithm for reinforcement learning though is Q-learning where agents try to learn the best strategy regarding a specific cost-function for possible actions that can be performed [10]. An exhaustive overview of applications is given in sources [8, 10–14].

As a cross-functional task, PPC is suitable for ML. The level of implementation is still low for two reasons: literature rarely considers customer, environmental, and human-in-the-loop aspects when linking ML to PPC. Recent applications seldom couple PPC to logistics as well as to design of products and processes. Two key pitfalls are identified in the implementation: the complexity of using Internet of Things to collect data and the difficulty of updating the ML model to adapt it to the manufacturing system changes [8]. By identifying correlations and recurrent patterns in the existing planning data with unsupervised learning, tacit planning knowledge can be revealed and reintegrated into new PPC workflows. Based on the classification and clustering of both product and process data incl. their linkages, an approach for the knowledge-based support of PPC was presented [15].

Table 5.1 Overview of applications and algorithms, adapted from [10]. *Note* RF-Random forest; SVM-Support vector machine; KNN-K-nearest neighbor; MLP-Multilayer perceptor; GBT-Gradient boosted trees; NLP-Neuro-linguistic programming

Subtopic	Application	Algorithm
Manufacturing process planning	Scheduling	Q-learning, RF, decision tree, KNN
	Cost and energy prediction	NN, SVD, GBT, RF, others
	System modeling	Log. Regr., RF, decision tree, Bayesian network
Quality control	Quality cost reduction	Decision tree, NN, SVM
	Process line quality	Decision tree, Bayesian network
Predictive maintenance	Remaining useful life	Decision tree, NN, PCA, KNN
Logistics	Scheduling	NN, Q-learning, deep q-learning, RF
Robotics	Human robot collaboration	Hidden Markov model, KNN, clustering, NN
	Path planning	KNN, NN
	Contact problem	Q-learning
Assistance and learning systems	Assembly assistance	NN
Process control & optimization	Production line	GBT, NLP
	Process and tool condition forecast	NN, trees, RF, SVM, mult. Regr
	Safety control	NN, SVM, Q-learning

In AM processing, modern ML algorithms can help to optimize process parameters, and examine powder spreading and in-process defect monitoring. On the production by AM, ML is able to assist practitioners in pre-manufacturing planning, and product quality assessment and control [16]. Concern about data security in AM as data breaches could occur with the aid of ML techniques is also addressed [17].

To obtain smooth trajectories in high-speed continuous motion of machine tools efficiently, a NN-based direct trajectory smoothing method is developed. A NN agent outputs servo commands directly based on the current tool path and running state in every cycle. To achieve direct control [18], motion feature and reward models were built, and reinforcement learning was used to train the NN parameters without additional experimental data. Due to its simple structure and low computational demands, it can easily be applied to real-time CNC systems [19].

An integrated solution of predictive model-based quality inspection in manufacturing is developed by using ML techniques and Edge Cloud Computing. A holistic approach was chosen comprising the target-oriented data acquisition and processing, modelling and model deployment as well as the implementation in the existing IT plant infrastructure. An industrial use case in surface mount technology (SMT) manufacturing is shown to underline the procedure and benefits of this method. The results show that by employing this method, inspection volumes can be reduced significantly and thus economic benefits can be generated [6].

For the generalization ability of the machinery fault diagnostics, the deep transfer learning consisting of both transfer learning and deep learning components was developed [12]. In-process tool condition forecasting is derived based on a DL method. A long short-term memory network is designed to forecast multiple flank wear values based on historical data. A residual CNN is built to enable in-process tool condition monitoring, using raw signals acquired during the machining process. So future flank wear values could be precisely forecasted during the machining [20].

Deep reinforcement learning (DRL) is an alternative approach for solving cloud manufacturing service composition issues. DRL as a model-free AI method enables a system to learn optimal service composition solutions through training, which can thus circumvent the afore-mentioned problems with meta-heuristics algorithms. A logistics-involved DRL-based service composition is proposed. A dueling Deep Q-Network (DQN) with prioritized replay is designed as the DRL algorithm [7].

Wide area for utilization of ML is in robotics [21]. Compared with obstacle avoidance in open environment, collision-free path planning for duct-enter task is often challenged by narrow and complex space inside ducts. Challenge mainly comes from two aspects: the definition of distances from robot to obstacles and the fusion of multiple data. Ideas underlying the success of human to handling this kind tasks, variable optimization strategies and learning, for one robust path planner are adapted. Proposed planner applies reinforcement learning skills to learn proper self-motion and achieves robust planning [22]. For achieving robust behavior, state-action planner is creatively designed with three especially designed strategies. After these creative designs, the planner has been trained with reinforcement learning skills. With the feedback of robot and environment state, proposed planner can choose proper optimization strategies, just like the human brain, for avoiding collision between robot body and target duct [23].

To reduce the uncertainty, a tolerance specification method is developed which uses out the information that affects geometric tolerances selection and use machine learning methods to generate tolerance specifications. The realization of tolerance specifications is changed from rule-driven to data-driven. This approach considers the past tolerance specification schemes as cases and sets up the cases to the tolerance specification database. The final results show that the machine learning algorithm can automatically generate tolerance specifications, and after feature engineering, the accuracy of the tolerance specification results is improved [24].

Wearable augmented reality (AR) smart glasses have been utilized in various applications such as training, maintenance, and collaboration. For sake of training, a smart and user-centric task assistance method is proposed, which combines deep learning-based object detection and instance segmentation with wearable AR technology to provide more effective visual guidance with less cognitive load. In particular, instance segmentation using the Mask R-CNN and markerless AR are combined to overlay the 3D spatial mapping of an actual object onto its surrounding real environment. 2.5D or 3D replicas support the 3D annotation and collaboration between different workers without predefined 3D models. Therefore, the user can perform more realistic manufacturing tasks in dynamic environments [25].

During the practice of the system upgrade, the redundancy of manufacturing resources and the inefficiency in resource configuration are the major obstacles to achieving satisfying value-creation within manufacturing ecosystems (ME). The interaction model provides a self-organized production mode without human intervention. The blindness, lag, and unfairness in the manual communication is eliminated by the Machine to Machine (M2M) interaction and automatic coordination. An NLP-based machine learning algorithm is introduced for quantifying semantic distance and measuring the differences between orders. Production resources can be scheduled and managed autonomously and the order-based production processes could be promoted seamlessly. An improved industrial ecosystem with automatic decision making and self-organization for future intelligent manufacturing is realized. The production cost is reduced by twelve percent the resource utilization rate improved and its economic value demonstrated [26]. A similar concept for trustable collaboration based on blockchain which utilizes customer views is shown in source [27].

The remainder of this chapter is structured as follows: in Sect. 5.2, fundamentals and problems concepts are illustrated. In Sect. 5.3, different gains of machine learning are discussed. Two use cases in plant engineering are described in Sect. 5.4. Data analysis is presented in Sect. 5.5, followed by discussion in Sect. 5.6. Finally, Sect. 5.7 summarizes the conclusions and outlook.

5.2 Fundamental and Problems Concepts

The current AI technologies, methods and procedures enable technical systems to perceive and process the environment and to solve problems on their own. The decision to solve problems is made from the learned information.

Machine learning is seen as a technological form of human ability in which the system analyzes situations, understands and makes decisions. Today's AI methods are not intended to copy human behavior, but AI technologies are considered to increase the effectiveness of individual processes. The important advantages of artificial intelligence are quality improvement [28], increase of process quality, time and cost saving. The more structured the processes are, the better the AI systems can be implemented, i.e. the current degree of autonomous decision-making depends heavily on how independently the system controls the complex situations within the defined system limits. The control of complex processes requires the knowledge of experience. For simple decisions like if-then-decisions or simple control loops in the field of automation technology no AI is needed. However, as soon as a solution is sought for complex situations in the solution space with multiple unknowns, AI can make an important contribution to solving the problem. The solution of a machine learning system is limited to the space of data known to the system. A good example

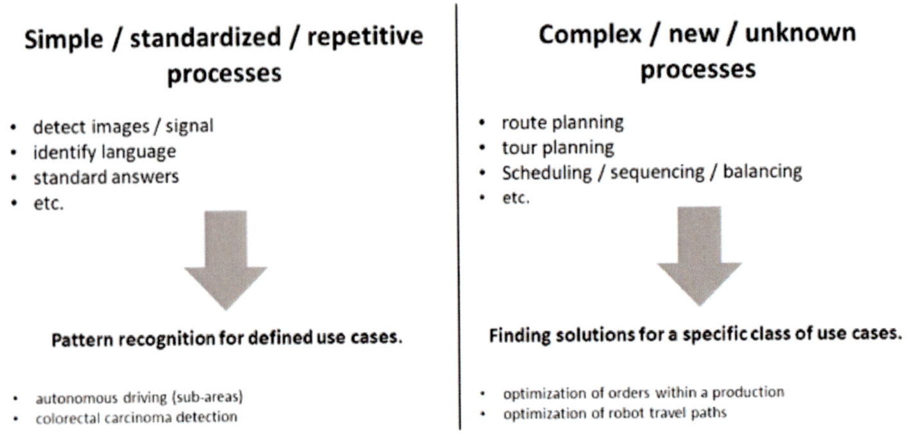

Fig. 5.1 Overview machine learning versus evolutionary algorithms

is the human being, as long as the child has not learned to walk upright, it cannot walk upright.

If a solution is sought outside the known data solution space (solution covered by the data), machine learning techniques are no longer the appropriate means to solve the problem. In this case, problem solving algorithms such as Monte Carlo Simulation, Evolutionary Algorithms can find better solutions. Figure 5.1 shows the main characteristics and comparison of problem solving based on ML resp. evolutionary algorithms.

5.2.1 When Is Machine Learning Used?

Machine learning is used whenever it is necessary to map standardized and repetitive processes. By recording the individual repetitions, one gets a recurring image of the same situation. In the last few years, neural networks have made great progress in the area of image recognition (classification/location of the object), as the database is becoming larger and larger and a situation can be viewed from multiple angles [29]. With the single view the AI system tries to recognize common features. The more features available, the better the prediction of the system [30]. The visual recognition of objects with AI is indispensable in the field of Industry 4.0 in many areas. An AI system can increase the recognition rate very well, especially for quality assurance in which, for example, components have to be inspected.

The implementation of AI systems in the area of structured data, especially the construction of Digital Twins based on measurement series or simulation data, requires an increased level of knowledge of the system that is to be represented by an AI system. Classically, these models are currently created with toolboxes such as MATLAB and DYMOLA. In an alternative case, the functions are modeled by using

Neural Networks	Evolutionary Algorithms
Modeled on the human brain.	Modeled on evolution: "Survival of the fittest".
Areas of application • pattern recognition • classification	**Areas of application** • optimization strategies
Necessary data • clean database assignment (input to output)	**Necessary data** • definition of the optimization goal based on the unknown
Advantage • constant computing time when processing the input data	**Advantage** • fluctuating computing time • scalable via spark platform
Disadvantage • lots of data sets necessary • solution space given by learning data	**Disadvantage** • validation more difficult due to the heuristic approach

Fig. 5.2 Detailed comparison machine learning versus evolutionary algorithms

the individual software on the basis of the mathematical models and relationships. The mathematical models then have to be validated in the subsequent step, which is a time-consuming process.

A more detailed comparison Machine Learning vs. Evolutionary Algorithms is illustrated in Fig. 5.2.

5.2.2 Where Are Evolutionary Algorithms Used?

For the solution of problems in the industrial environment, solution approaches such as evolutionary algorithms or genetic algorithms, which are based on evolution, are suitable. Here is an example from nature. Today we know the hare with long ears and as a very fast and nimble creature. That was not always so. In the past, the brown hare had short ears and not so strong hind legs. What happened? The hares with short ears could not hear the enemy and were eaten by him. The hares with the weak hind legs could hear their enemy with the big ears, but could not run away because they were too slow. For thousands of years the hare got bigger ears and stronger hind legs. Unfortunately, all other hares no longer exist. What has this now to do with our algorithm. With the algorithm of the evolutionary algorithms exactly this paradigm is represented. The individual steps are listed in Fig. 5.3.

An important difference to the classical solution approaches is that we do not try to calculate the solution. Instead, we generate any number of solutions and then evaluate these solutions. The procedure is:

1. Current accumulation of solutions. All solutions are called population.
2. Selection of the based on the previous evaluation. In the first run, there is no selection because the solutions have not yet been evaluated.

Fig. 5.3 Steps of the evolutionary algorithm

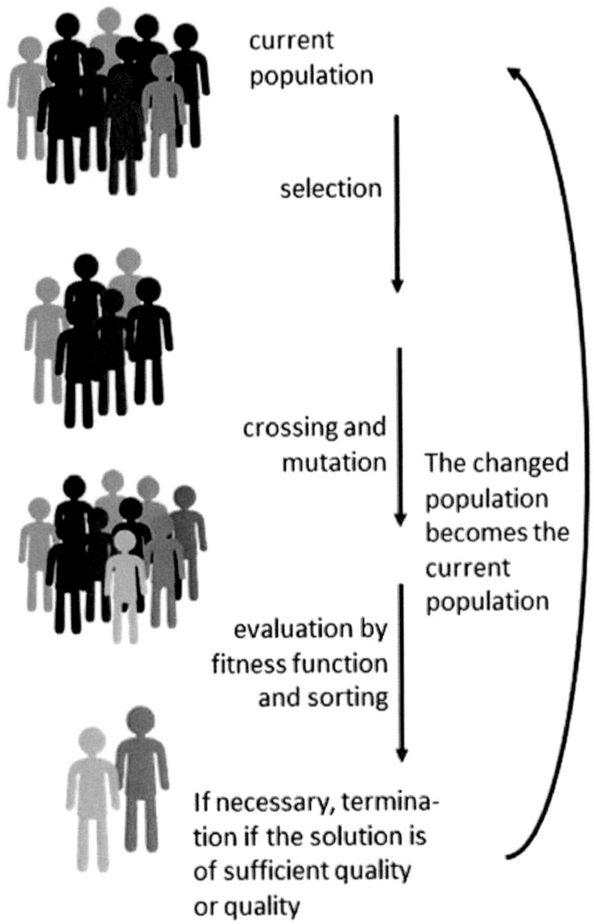

3. Generation of new solution from base crossing and mutation of the still existing solutions with each other. The deleted solutions are replenished by the new solutions.
4. Evaluation of the solution with respect to the achievement of the goal by means of a fitness function. The fitness function describes the goal achievement of the solution with respect to the optimization goal.
5. Deletion of bad solutions based on the result of the fitness function.

Bad solutions are deleted again. Good solutions are retained. The mutation and selection also save the last good actions for mutation and selection and executes them again at the next step if necessary. With this procedure, the quality of the solution slowly improves from loop pass to loop pass. To avoid the problem of the quasi-optimal optimization point, completely new solutions are repeatedly created and tested in the individual loops.

This type of problem solving is useful when I know my data set very well and the solution has a lot of unknown quantities.

5.2.3 A Combination of Several Methods of AI?

Especially, in the area of control of complex systems, there is no "AI system" that solves all problems. In this case, a combination of several approaches from the field of AI is used. For example, it is necessary to optimize the trajectory of several robots in a welding cell. The problem is to solve multidimensional space (axes of the robot, displacement on the floor, location, etc.). The first step is to create a Digital Twin for each robot. This Digital Twin describes the behavior of the robot in different situations as a function of time. The models of the individual robots are then used by the optimization algorithm and the trajectory is optimized according to the defined rules.

The following Fig. 5.4 shows the energy consumption of a 3-axis robot as a function of speed. The faster the robot travels, the higher the initial energy consumption i.e. in the context of a CO_2 neutral factory, energy peaks must be largely avoided in the process, as this energy must be provided by large batteries or by the electricity supplier.

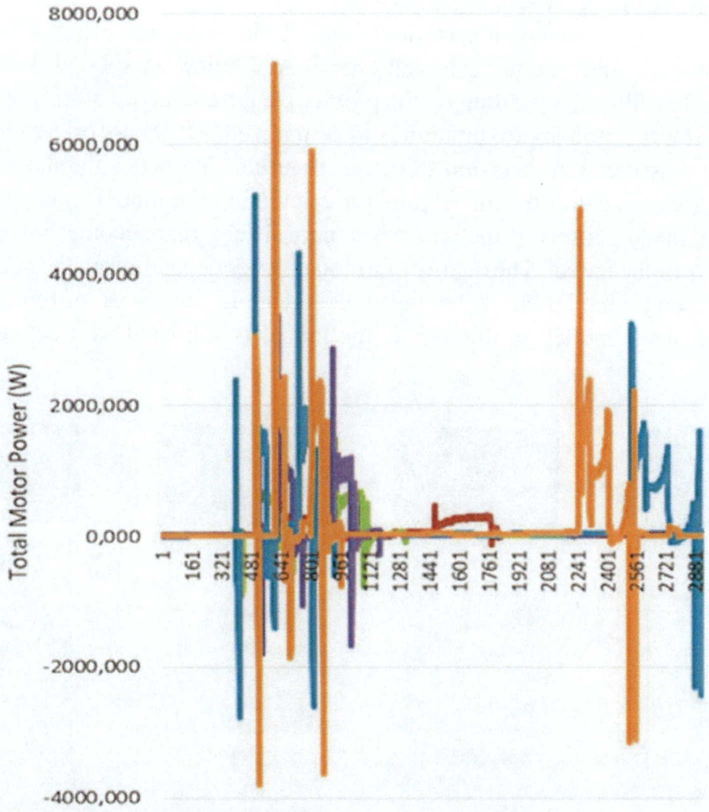

Fig. 5.4 Energy consumption of a 3-axis robot

5.2.4 Procedure for Creating a Behavioral Model/Digital Twin

The creation of a Digital Twin on the basis of real data or simulation data is carried out in the following steps (Fig. 5.5):

1. **Selection of the data**
   The selection of data depends on the goal of the Digital Twin. If there are several data sources to be used, clear references are required for correlating the data with each other.
2. **Preprocessing of the data**
   With the pre-processing of the data, the selected data is analyzed and searched for recurring patterns or areas of particular importance. For the search, the data can be visualized with different types of charts. Another way is to calculate metrics. However, at the beginning it is recommended to visualize the data, because humans can recognize patterns better visually. At the end of the step, there is a consistent data set, which is used for model building.
3. **Create behavior model/Digital Twin**
   In this step, the behavioral model/Digital Twin is created using AI methods. The model is created using a well-known AI framework such as TensorFlow, Pytorch, Coffee. Depending on the problem, a different type of network is used [31]. If a function approximation is to be realized, MLP networks can be used. For a classification, decision trees can be used. The better the data has been prepared, the better the algorithms for calculating the model can approximate the learned data sets. If the data set is inconsistent, that becomes visible in the quality of the model. The quality of a model varies depending on the application. How good a model has to be, becomes visible in the course of the validation phase of the model. In this phase, the model is validated with actual data. If

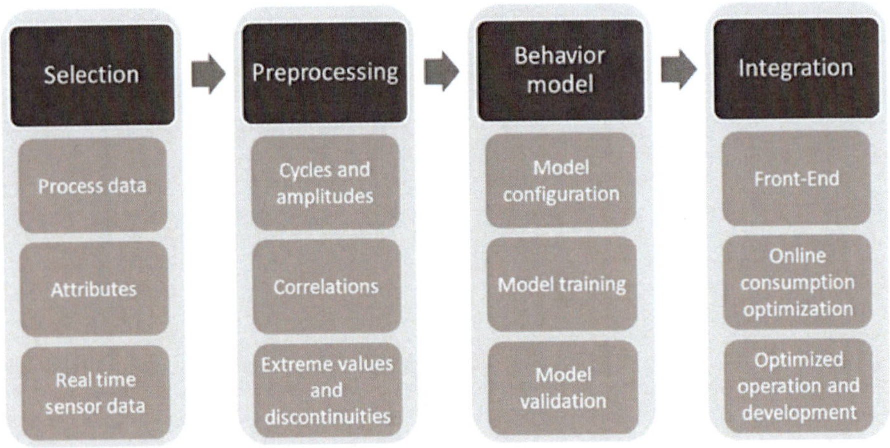

Fig. 5.5 Principle schema of the workflow of the AI system to be developed

the quality is too poor, the data must be reprocessed again in step 2 with the new findings. Finally, the validation and quality check of the model can be done on the remaining twenty percent of the customer's raw data. In this step, too, a-priori knowledge is introduced into the process.

4. **Integration**
   In the last step, the model is deployed in the productive environment. As a rule, the model gets an interface in the existing system environment. The system, which uses the model, comes to a large extent from the specific characteristics of the open-source framework like TensorFlow, etc. nothing with. During the deployment process is to be examined the reaction time of the entire system. Especially, if the model is integrated into a real-time control, the values have to be checked carefully. However, the more hidden layers the model has, the more computing power the model needs to calculate the values. If the model is set productive, validations of the prediction quality of the model with the current data stock are to be accomplished from time to time. If necessary, the model must be further trained with new data sets.

By means of correlation analyses as well as course comparisons between real data and model data, possible optimizations of the physical system can be concluded, in particular, when observing close to the limits of the working range: Saturation effects in individual parameters or in the course of the time series or oscillation phenomena indicate further optimization potentials.

5.3 Gains of Machine Learning

Many advantages can be identified for machine learning in the context of Industry 4.0. However, one thing is clear: the following advantages can only be used if data from production is already available in the necessary quality, quantity and consistency [32].

5.3.1 Efficient Derivation of the Model Configuration Based on the Data

The structure of the model is derived on the basis of the data analysis. For this purpose, different types of models can be used depending on the application.

For a model based on a neural network, several parameters are responsible for an optimal network (Table 5.2).

Only when all parameters are tuned, an optimal mesh of the desired quality is created. For this purpose, the suitability of different classification approaches for the selection of an optimal parameter combinatorics for the representation of the

Table 5.2 Important parameters for configuring networks

Parameter	First indications of the values of the parameters
Number of layers	• Findings based on the correlation of parameters with each other • Number of data sets
Number of neurons per layer/structure of the layers	• Findings based on the correlation of the parameters to each other • Adjustment based on the analysis of discontinuities
Activation functions (Relu, tanh, Linear, etc.)	• Adaptation based on the analysis of discontinuities and the necessary run-time behavior of the model
Optimization algorithm for calculation of weights	• Fitting based on data structure, extreme value analysis and number of degrees of freedom • Evaluation function to evaluate the quality of the result of the optimization algorithm

dynamics underlying the original system must be investigated or system cascade must be selected during the development.

For the modeling not always a kind of the neural network is necessary. For example, cascaded decision trees or the definition of a mathematical assignment rule using the data structures on the basis of suitable criteria are often more useful if the data in the analysis step produce a clear correlation.

5.3.2 *Generating Behavioral Model Through Training with Data*

With the neural network, the behavior of the system is mapped virtually. By training with the pre-processed input data, the system is enabled, for example, to permanently approximate the course of functional physical dependencies.

In order to optimize convergence, learning rules such as Hebb rule, delta rule, backpropagation, competitive learning, etc.—which always have certain advantages and disadvantages depending on the concrete data structure and the underlying model dynamics—must be selected on the basis of the preliminary data analysis and combined with each other if necessary. Here, too, correlation analyses and the identification of functional relationships between data structures and the suitability of the learning rules must be used to define the optimal configuration of the system training without specific data and information technology knowledge.

The neural network is trained on the basis of established 80–20 approaches. The suitability of further nonlinear optimization functions must be checked in each individual case and must be included in the network configuration (Fig. 5.6).

In the following, the parameters that have a significant influence in combination with the parameters from the previous chapter are filled in (Table 5.3).

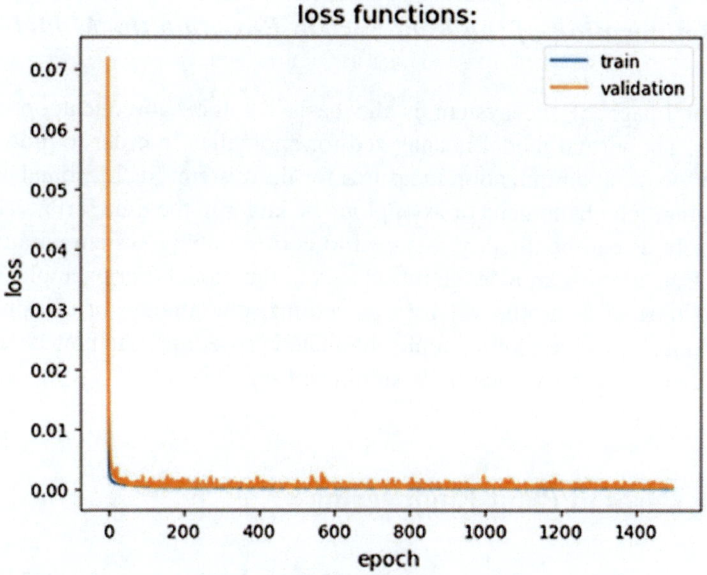

Fig. 5.6 Optimal course of a learning curve in TensorFlow

Table 5.3 Important parameters for teaching nets

Parameter	First hints
Batch Size	• Adjustment based on the speed of convergence across training epochs
Loss function	• Based on the learning progression across the eras visible
Optimization function	• Analysis of the balance of the calculated weights of the first layers of a mesh

5.3.3 Easy Validation of the Model

Due to the division of the dataset into 80% learning data and twenty percent validation data, an accuracy is calculated for the twenty percent of the validation data. The deviation of the measured values compared to the calculated values is used as a quality characteristic. Here the quality of several data must be checked and evaluated automatically, in order to ensure the durable suitability of the model for the reproduction. A good model can predict all new/unknown values in the same way. The accuracy over the new data remains constant [33].

If the model has the necessary quality, the so-called test operation begins, in which the system is initially integrated silently into the target system. Before the validation period, the optimal weighting of the individual influencing factors, as well as the mathematical structure of the accuracy cost function (quadratic measure, non-linear function based on Lagrange operators…) must be defined for the correct evaluation of the quality in the running operation.

5.3.4 Optimization of the Real System Based on the Model

This virtual image of the system is the basis for the subsequent optimization approaches. The virtual model is analyzed for anomalies in order to draw conclusions about possible optimization measures for the system. Such anomalies are for example saturation phenomena or asymptotic behavior in the time series. The mathematic structures can be directly assigned to corresponding system parameters on the basis of the a priori knowledge introduced in the model. For example, to adjust the closing time of a throttle valve or to optimize the amount of compressed air introduced into a cylinder. Very simple automated procedures can now be used with the digital twin to identify potentially significant sections.

5.4 Use Cases in Plant Engineering

In the following sections, two exemplary use cases from our practice are examined in more detail.

5.4.1 Mechanical Model of a Pneumatic Cylinder

A pneumatic cylinder is mostly used when a translation is to be performed, such as sorting goods on a belt into different containers. The travel speed of the cylinder, depending on how much weight the cylinder has to move, correlates with the set opening angle of the throttle valves for the two directions of movement (extension and retraction). If the opening angle is opened too wide, the cylinder moves quickly to its destination and a lot of compressed air is required. If the cylinder moves too slowly, the process to be performed may not be carried out correctly. The cylinder consumes little compressed air. Until now, the adjuster has largely set the opening angle based on his feeling. As a rule, more compressed air was used here than was necessary. With an AI system that maps the position and travel speed, it is possible to optimally set the opening angle in the context of compressed air consumption and process speed. With this approach, the cylinder is operated in the optimal range by the adjustment and the lifetime is increased. If continuous monitoring of the component is required, the model can be run as a digital shadow and a continuous comparison between target and actual can be performed. If the target/actual comparison leaves a certain tolerance band, an automatic check of the related component is initiated. The model can be used in different simulation environments. If the model is used to optimize compressed air consumption, the compressed air consumption is reduced by at least ten percent (Fig. 5.7).

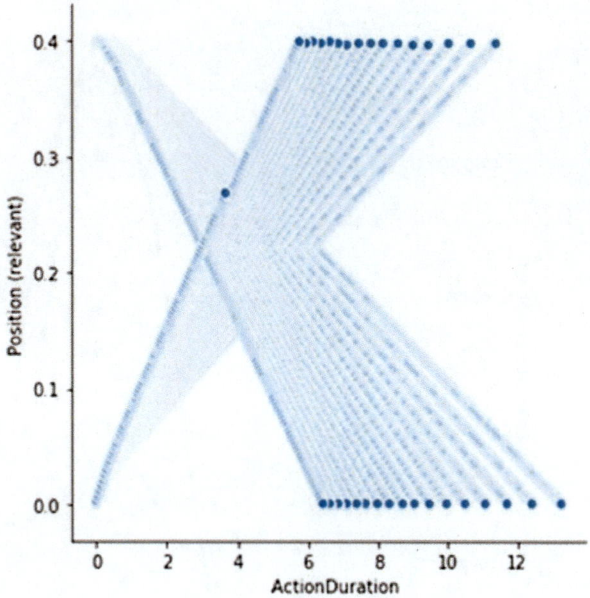

Fig. 5.7 Course of a pneumatic cylinder at different opening degrees

5.4.2 Forecast Models Solar System

In the course of the climate neutrality of factories, the own generation of electricity with a solar plant and its prediction of the generated power in connection with the irradiation of the sun is indispensable. In the case of irradiation, the direct radiation, which radiates directly onto the solar system, and indirect radiation, which is emitted by the environment, are relevant for the calculation of the power output. Existing mathematical models describe the direct irradiation very well. However, as soon as the models have to predict diffuse irradiation, the environment in which the solar system is installed becomes very important. In this case, models based on measured data are twenty percent superior to mathematical models. A good side effect of the model based on AI is that as soon as the performance of the solar system is no longer correct due to overgrowth of e.g. grass/dirt, this fact becomes immediately visible in the target/actual comparison. Finally, the predictive maintenance is included in this model [34]. If the complete prediction chain of weather over solar system is calculated, an improvement of ten percent with diffuse light comes out. With the improved forecast, a significant contribution is made to the climate change in the field of electricity generation (Fig. 5.8).

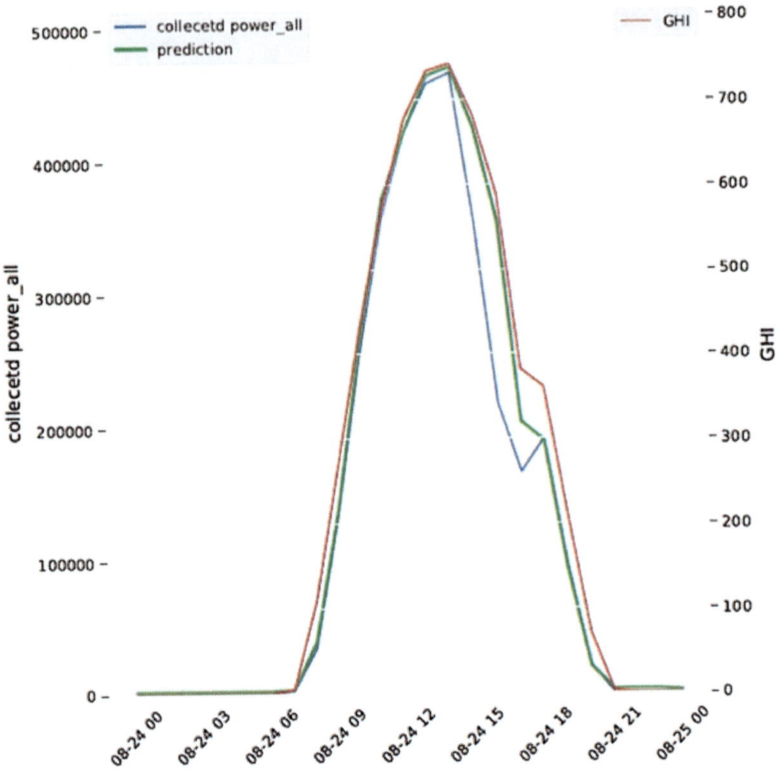

Fig. 5.8 Course of a solar radiation (red) in relation to real (blue) and estimated (green) emitted power

5.5 Data Analysis

Data analysis is the process of obtaining information from data. First of all, it is directly about information about the available data stock, but based on this, knowledge about the data stock can usually also be derived.

In a real data science project, there is usually a large amount of data available and it is necessary to examine it with regard to a specific question in order to make reliable statements on the basis of the data. First of all, however, the completeness and correctness of the data itself must be doubted; data sets are generally not to be regarded as trustworthy until they themselves have proven a high degree of consistency in the analysis. Typical sources of error for faulty data are, for example, defective or non-comparable sensors e.g. a sensor change can cause an offset shift. But also changes in the underlying metrics when calculating derived values and especially human errors when compiling data sets. There is a ML-based technique for data-driven fault detection to online handle nonlinear dynamic systems e.g. chemical process [35].

In the initial analysis of a data set, each value series is first considered separately. Only then are correlations between the measured variables examined. Using the measures of descriptive statistics, initial statements can be made quickly and efficiently about the individual measured variables, such as the mean value and the quartiles. Are minimum and maximum in the physically possible range at all or do these indicate invalid values? Which statements do the moments of the curve make about the distributions?

A human strength is the processing of visual stimuli, many methods of explorative statistics, finding correlations in data are based on the interpretation of graphs. For individual measured variables, histograms or box plots provide compact information about the underlying relationships in a clear manner. Even with a quick look at the beginning and end of the data set at the same time, inconsistencies can sometimes be recognized immediately.

In practice, the examination of data quality almost always reveals conspicuities (as explored in Chap. 7). Data analysis projects should always be carried out in close cooperation with the relevant departments. It often makes sense to take measures based on the examination of individual measured variables, such as defining threshold values for valid measurement ranges. Surprisingly often, even the simplest analyses reveal points that were previously overlooked and deliver direct added value, quick wins.

Sources of error also lie in information technology processing. For example, non-existent values are mapped differently in different programming languages. The correct distinction between the physical value zero and a non-existent value must be checked in each case. The distribution of the existing values to the blanks gives information about the data availability. This can have serious influence on the data analysis, if a sensor becomes "blind" under unfavorable environmental conditions [36], it will systematically exclude critical measuring ranges. In addition to non-existing values, the data types and the type conversions, which are sometimes performed automatically by programs, are also a source of error. If these are not recognized at the input and for example a part of the data receives another data type, these are not considered with further computations. This leads to systematic distortions.

Once the basic work to ensure the general data quality for the individual measured variables and their quantity has been done, comes the exciting part—finding correlations in the data. Fast and efficient is the mapping in a so-called pair- or scatterplot matrix. In this matrix, two variables are always plotted in pairs against each other, correlations are often already visible in the first plot. In some cases, it is helpful to consider only a part of the measured variables, also the data can be thinned out by randomized sampling for better clarity. Color and type of markers are further means to represent the information of more than two variables in one subgraph. Despite the possibilities to condense the information, it makes sense to check all possible correlations in an unbiased way and to compress them only later, for example to communicate the final results. Data analysis should judge un-biased on the basis of the data, the danger of confirmation bias should not be underestimated.

While the scatterplot matrix depicts the visual aspect of finding correlations, this can be supported by specific statistical key figures such as the correlation coefficient.

Once correlations are identified in the data, hypotheses can be neatly proven using statistical tests. In practice, however, discussion with subject matter experts is often useful and more goal-oriented at this point. As a rule, the basis for models to explain the data can optimally be determined in joint expertise on the one hand from the point of view of the findings of the data analysis developed so far and on the other hand from the experience of the respective discipline. These behavioral models can be used in the next steps to answer questions, for example, to create assistance [25] or predictive maintenance [30] systems (Fig. 5.9).

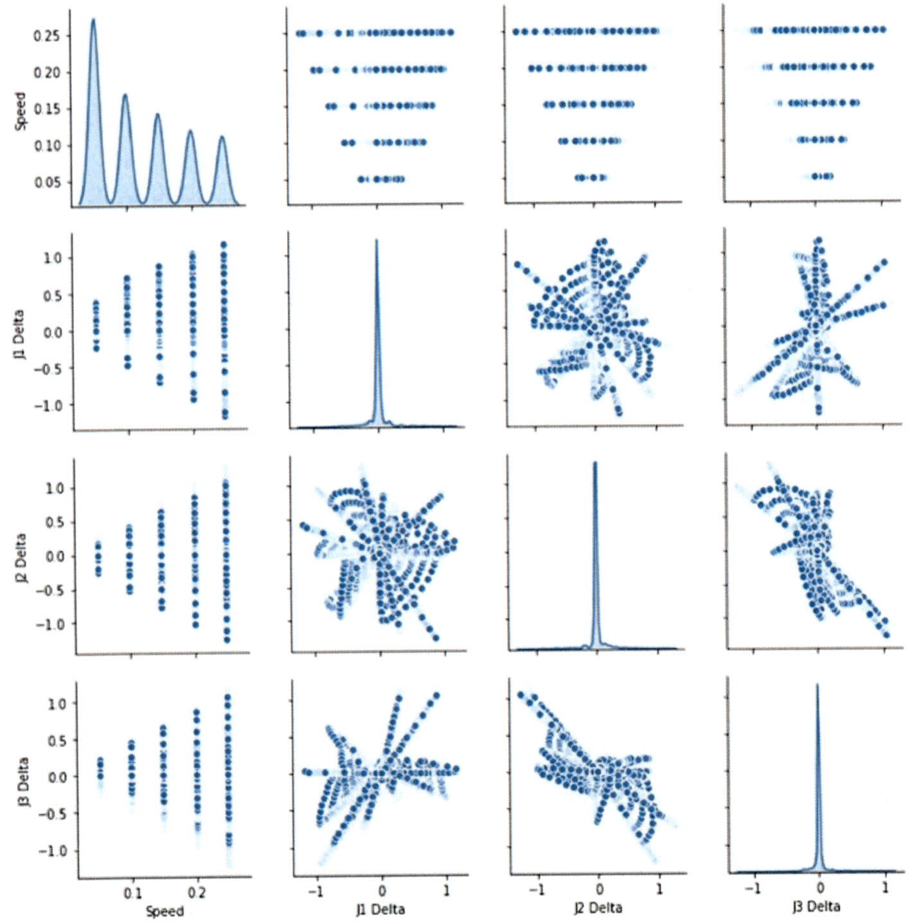

Fig. 5.9 Pair plots of an axis robot

5.6 Discussion

Artificial intelligence (AI) is becoming increasingly important in all areas of human life. In the areas of voice control and chatbots, the use of AI is already indispensable [2]. A big driver was the exploitation of existing data in the areas via social media and the research of recent years in the environment [37]. For the classification of images (recognition of texts, animals on pictures), the recognition rates are getting better and better [38, 39]. Especially in the area of assistance systems for supporting diagnoses in the medical field, the quality of the systems has also increased massively in recent years [40].

In the area of manufacturing, first productive implementations have already been successfully carried out. However, the topic of predictive maintenance is still waiting for its broad breakthrough. The data basis for machine learning is only just being established. The question that always arises here is what data and at what measurement speed are necessary? The type of data depends on the question to be answered. For a sufficient model, data from the following areas in a sufficient extent are required:

- Process data: The data generated by the manufacturing process, e.g. temperature profile at a certain point in a press mold.
- Product data: The data describing the property of the input product(s) into the process, e.g. viscosity of rubber in tire production.
- Plant data: Data describing the setting of the process in the equipment. In the case of resistance welding, how high is the maximum current readjusted.

The above-mentioned advantages of ML are not intended to cover the pitfalls which can occur primarily by use of not appropriate data [41]:

- Data Leakage: The problem of making use of data in the model to which a production system would not have access e.g. in time series problems.
- Overfitting: Modeling the training data too closely and too specific causing the low ability to generalize. This becomes more of a problem in higher dimensions with more complex class boundaries.
- Data Sampling and Splitting: Train/test/validation sets must be independent samples.
- Data Quality: Inconsistent, duplicate, and corrupt data needs to be identified and explicitly handled (Chap. 7). It can contaminate the modeling problem and ability of a model to generalize [42].

According to expectations, Digital Twins created with AI methods are on average twenty percent more accurate than classical models and require only 50 percent of the development time. In a further step, the models can be used for any application, such as validation of optimization strategies [43], the digital shadow [44], elimination of sensors [45], which can be generated via the model, and much more.

5.7 Conclusions and Outlook

This chapter provides a comprehensive insight into approaches for industrial Implementation of Machine Learning, in particular Deep Learning. As compared with simple neural networks, deep neural networks are more robust and efficient in representing complex function and detecting features of the given input data [37]. Various models and procedures can be trained using more than one method, so their efficiency depends upon the domain in which they are used and the quality of the training data used. Many open-source data sets are available as the training base for different domains [11]. Hence the quality of the data set is a huge concern (Chap. 7). A brief overview of the few subtopics and applications from manufacturing where ML can be used is presented in this chapter. ML can be used in almost any domain of manufacturing, but the cost of the sophisticated hardware, in particular for the training, is the inherent obstacle [37].

In the future, the creation of Digital Twins with AI methods will become prevalent in order to achieve the goals set in connection with efficiency, autonomy, sustainability. To achieve the next big advance, further developments of ML methods are necessary [37]:

1. The current networks produce a single final prediction for the input. For example, an image is classified into either yes or no for a category. But sometimes, the image is actually very confusing or damaged. In these cases, a network should produce an entire set of possible outcomes.
2. Networks are trained on thousands of images to detect objects from images. Current networks work on the structural properties of the objects. If generative models are trained on the physical construction properties on objects, the results will be much better.
3. Capsule architectures are the future of deep learning. The capsule architecture is a better way for routing information from lower layers to higher layers.
4. Generative models such as deep convolutional generative adversarial networks, these models are learning how the world actually looks like instead of memorizing the examples.
5. As a mid-term vision of development, progressive neural networks are able to learn continually by accumulating knowledge and transferring them to a new domain. They are immune to catastrophic forgetting and can leverage previously gathered knowledge from pre-trained models via lateral connections and continually learn new tasks and thus has been shown to perform well in reinforcement learning tasks [11].

Machine builders and operators use ML solutions to optimize their production and to offer new data-based services. In order to identify the relevant relationships in the data, these are analyzed using machine learning methods and models are developed for anomaly detection, classification or error prediction, among other things. However, real-world approaches are rare to get the benefit of some of their lessons and best practices. Taulli [46] presents one of them which starts with building of an implementation team. Data preparation as the first technical task is absolutely critical providing strategies to improve the results. Subsequently, key steps are presented for building an effective algorithm using the various AI platforms that help to streamline the process. The deployment and monitoring parts of the AI process provide ways to minimize the risks. Finally, handling of bias in models is also addressed.

Like other transdisciplinary approaches, Machine Learning has a societal impact too and causes some ethical concerns [47]. ML, in particular Deep Learning, is often compared with mechanisms that underlie the human mind, and some scholars believe that it will continue to rise at an unexpected growth and attract many more domains. Fear arises that deep learning might threaten the very social and economic structures that hold our societies together, by either driving humans into unemployment or slavery. There is no doubt that ML and DL are super-efficient for many tasks. In contrast, that are not universal techniques (even DL) which can solve all problems and override all previous technologies. This learning technique still has limits and challenges which prevent it from competing with the mind of a human being. Human beings can learn abstract, broad relationships between different concepts and make decisions with little information [48].

Such ethical concerns can be resolved by further research by inspiration drawn from cognitive neuroscience and developmental psychology. Deciphering human behavioral patterns may facilitate major breakthroughs in applications such as enabling artificial agents to learn about spatial navigation on their own which comes naturally to most living beings [11]. In terms of theoretical contribution, two different deep learning models can be integrated that can simultaneously provide two challenging tasks e.g. recognize the action of an operator and detect the object. In this way, skill transfer in manufacturing systems can be facilitated by adapting or learning new skills for junior operators in a workshop [49].

Similarly, a dynamic clustering model can be designed for fault diagnosis in a workshop using the proposed semi-supervised multi-spatial manifold clustering method to recognize attribute clusters and aggregate new types. When new types are added to this model, it is constantly updated to achieve the automatic evolution of the knowledge base for the diversity of fault. A knowledge evolution model is established by the generative adversarial network algorithm to achieve self-learning and self-optimizing capabilities of the knowledge base. This method generates new knowledge elements to optimize the clustering model [50]. It supports theories from neuroscience that establish grid cells as a critical component for vector-based navigation. The latter can be coupled with path-based strategies to aid in navigation in challenging environments. Lacking scientific explanations of how humans solve complex visual and auditory problems, a section of cognitive scientists have embraced deep neural networks as models of human brain responses and behavior [11]. First of all, such advances could improve the human–machine interface [51, 52]. Further application reflects business processes reconstruction in order to systematically and fully consider the quality correlations in finer-grain, i.e. task-level or activity-level, and temporal performance in a process model optimization [53].

References

1. Haenlein M, Kaplan A, Tan CW, Zhang P (2019) Artificial intelligence (AI) and management analytics. J Manage Anal 6(4):341–343. https://doi.org/10.1080/23270012.2019.1699876
2. Lu Y (2019) Artificial intelligence: a survey on evolution, models, applications and future trends. J Manage Anal 6(1):1–29. https://doi.org/10.1080/23270012.2019.1570365
3. Trappenberg TT (2020) Fundamentals of machine learning. Oxford University Press, Oxford. https://doi.org/10.1093/oso/9780198828044.001.0001
4. Sajja PS (2021) Illustrated computational intelligence: examples and applications. Springer, Singapore. https://doi.org/10.1007/978-981-15-9589-9
5. Sze V, Chen YH, Yang TJ, Emer JS (2020) Efficient processing of deep neural networks. Morgan & Claypool, San Rafael. https://doi.org/10.2200/S01004ED1V01Y202004CAC050
6. Schmitt J, Bönig J, Borggräfe T, Beitinger G, Deuse, J (2020) Predictive model-based quality inspection using machine learning and edge cloud computing. Adv Eng Inform 45:101101. https://doi.org/10.1016/j.aei.2020.101101
7. Liang H, Wen X, Liu Y, Zhang H, Zhang L, Wang L (2020) Logistics-involved QoS-aware service composition in cloud manufacturing with deep reinforcement learning. Robot Comput Integr Manuf 67:101991. https://doi.org/10.1016/j.rcim.2020.101991
8. Usuga Cadavid JP, Lamouri S, Grabot B, Pellerin R, Fortin A (2020) Machine learning applied in production planning and control: a state-of-the-art in the era of industry 4.0. J Intell Manuf. https://doi.org/10.1007/s10845-019-01531-7
9. Hennig M, Grafinger M, Hofmann R, Gerhard D, Dumss S, Rosenberger P (2021) Introduction of a time series machine learning methodology for the application in a production system. Adv Eng Inform 47:101197. https://doi.org/10.1016/j.aei.2020.101197
10. Fahle S, Prinz C, Kuhlenkötter B (2020) Systematic review on machine learning (ML) methods for manufacturing new methodology to analyze the functional and physical architecture. Procedia CIRP 93:413–418. https://doi.org/10.1016/j.procir.2020.04.109
11. Sengupta S, Basak S, Saikia P, Paul S, Tsalavoutis V, Atiah F, Ravi V, Peters A (2020) A review of deep learning with special emphasis on architectures, applications and recent trends. Knowl Based Syst 194:105596. https://doi.org/10.1016/j.knosys.2020.105596
12. Li C, Zhang S, Qin Y, Estupinan E (2020) A systematic review of deep transfer learning for machinery fault diagnosis. Neurocomputing 407:121–135. https://doi.org/10.1016/j.neucom.2020.04.045
13. Zhong RY, Ge W (2018) Internet of things enabled manufacturing: a review. Int J Agile Syst Manage 11(2):126–154. https://doi.org/10.1504/ijasm.2018.10013695
14. Peruzzini M, Stjepandić J (2018) Editorial to the special issue "Transdisciplinary analytics in supply chain management." J Manage Anal 5(2):75–80. https://doi.org/10.1080/23270012.2018.1443405
15. Wallis R, Stjepandic J, Rulhoff S, Stromberger F, Deuse J (2014) Intelligent utilization of digital manufacturing data in modern product emergence processes. Adv Transdisciplinary Eng 261–270. https://doi.org/10.3233/978-1-61499-440-4-261
16. Wang C, Tan XP, Tor SB, Lim CS (2020) Machine learning in additive manufacturing: state-of-the-art and perspectives. Additive Manuf 36:101538. https://doi.org/10.1016/j.addma.2020.101538
17. Engelmann F, Holland M, Nigischer C, Stjepandić J (2018) Intellectual property protection and licensing of 3d print with blockchain technology. Adv Transdisciplinary Eng 7:103–112. https://doi.org/10.3233/978-1-61499-898-3-103
18. Lv W, Xiong J, Shi J, Huang Y, Qin S (2020) A deep convolution generative adversarial networks based fuzzing framework for industry control protocols. J Intell Manuf. https://doi.org/10.1007/s10845-020-01584-z
19. Li B, Zhang H, Ye P, Wang J (2020) Trajectory smoothing method using reinforcement learning for computer numerical control machine tools. Robot Comput Integr Manuf 61:101847. https://doi.org/10.1016/j.rcim.2019.101847

20. Sun H, Zhang J, Mo R, Zhang X (2020) In-process tool condition forecasting based on a deep learning method. Robot Comput Integr Manuf 64:101924. https://doi.org/10.1016/j.rcim.2019.101924
21. Amorim A, Guimares D, Mendona T, Neto P, Costa P, Moreira AP (2021) Robust human position estimation in cooperative robotic cells. Robot Comput Integr Manuf 67:102035. https://doi.org/10.1016/j.rcim.2020.102035
22. Winder P (2021) Reinforcement learning. O'Reilly Media, Sebastopol
23. Hua X, Wang G, Xu J, Chen K (2020) Reinforcement learning-based collision-free path planner for redundant robot in narrow duct. J Intell Manuf. https://doi.org/10.1007/s10845-020-01582-1
24. Cui LJ, Sun MY, Cao YL, Zhao QJ, Wen-han Zeng WH, Guo SR (2021) A novel tolerance geometric method based on machine learning. J Intell Manuf. https://doi.org/10.1007/s10845-020-01706-7
25. Park KB, Kim M, Choi SH, Lee JY (2020) Deep learning-based smart task assistance in wearable augmented reality. Robot Comput Integr Manuf 63:101887. https://doi.org/10.1016/j.rcim.2019.101887
26. Li J, Zhang Y, Qian C, Ma S, Zhang G (2021) Research on recommendation and interaction strategies based on resource similarity in the manufacturing ecosystem. Adv Eng Inform 46:10118. https://doi.org/10.1016/j.aei.2020.101183
27. Barenji AV, Guo H, Wang Y, Li Z, Rong Y (2021) Toward blockchain and fog computing collaborative design and manufacturing platform: support customer view. Robot Comput Integr Manuf 67:102043. https://doi.org/10.1016/j.rcim.2020.102043
28. Li X, Jia X, Yang Q, Lee J (2020) Quality analysis in metal additive manufacturing with deep learning. J Intell Manuf 31:2003–2017. https://doi.org/10.1007/s10845-020-01549-2
29. Chouai M, Merah M, Sancho-Gómez JL, Mimi M (2020) Supervised feature learning by adversarial autoencoder approach for object classification in dual X-ray image of luggage. J Intell Manuf 31:1101–1112. https://doi.org/10.1007/s10845-019-01498-5
30. Kim TS, Sohn SY (2020) Multitask learning for health condition identification and remaining useful life prediction: deep convolutional neural network approach. J Intell Manuf. https://doi.org/10.1007/s10845-020-01630-w
31. Ktari A, El Mansori M (2020) Digital twin of functional gating system in 3D printed molds for sand casting using a neural network. J Intell Manuf. https://doi.org/10.1007/s10845-020-01699-3
32. Kuhnle A, Kaiser JP, Theiß F, Stricker N, Lanza G (2020) Designing an adaptive production control system using reinforcement learning. J Intell Manuf. https://doi.org/10.1007/s10845-020-01612-y
33. Bustillo A, Reis R, Machado AR, Pimenov DY (2020) Improving the accuracy of machine-learning models with data from machine test repetitions. J Intell Manuf. https://doi.org/10.1007/s10845-020-01661-3
34. Alelaumi S, Khader N, He J, Lam S, Yoon SW (2021) Residue buildup predictive modeling for stencil cleaning profile decision-making using recurrent neural network. Robot Comput Integr Manuf 68:102041. https://doi.org/10.1016/j.rcim.2020.102041
35. Said M, ben Abdellafou K, Taouali O (2020) Machine learning technique for data-driven fault detection of nonlinear processes. J Intell Manuf 31:865–884. https://doi.org/10.1007/s10845-019-01483-y
36. Forte G, Alberini F, Simmons M, Stitt HE (2020) Use of acoustic emission in combination with machine learning: monitoring of gas–liquid mixing in stirred tanks. J Intell Manuf. https://doi.org/10.1007/s10845-020-01611-z
37. Aditya Khamparia A, Singh KM (2019) A systematic review on deep learning architectures and applications. Exp Syst 36:e12400. https://doi.org/10.1111/exsy.12400
38. Liu Li, Ouyang W, Wang X, Fieguth P, Chen J, Liu X, Pietikäinen M (2020) Deep learning for generic object detection: a survey. Int J Comput Vision 128:261–318. https://doi.org/10.1007/s11263-019-01247-4

39. Salem B, Stjepandić J, Stobrawa S (2019) Assessment of methods for industrial indoor object recognition. Adv Transdisciplinary Eng 10:390–399. https://doi.org/10.3233/ATDE190145
40. Yang S, Niu J, Wu J, Wang Y, Liu X, Li Q (2021) Automatic ultrasound image report generation with adaptive multimodal attention mechanism. Neurocomputing 427:40–49. https://doi.org/10.1016/j.neucom.2020.09.084
41. Clemmedsson E (2018) Identifying pitfalls in machine learning implementation projects: a case study of four technology-intensive organizations. Master thesis, KTH Industrial Engineering and Management Industrial Management
42. Gürdür D, El-khoury J, Nyberg M (2019) Methodology for linked enterprise data quality assessment through information visualizations. J Indus Inform Integr 15:191–200. https://doi.org/10.1016/j.jii.2018.11.002
43. Cheng Y, Chen K, Sun H, Zhang Y, Tao F (2018) Data and knowledge mining with big data towards smart production. J Indus Inform Integr 9(2018):1–13. https://doi.org/10.1016/j.jii.2017.08.001
44. Ladj A, Wang Z, Meski O, Belkadi F, Ritou M, Da Cunha C (2020) A knowledge-based Digital Shadow for machining industry in a Digital Twin perspective. J Manuf Syst. https://doi.org/10.1016/j.jmsy.2020.07.018
45. Li W, Wang X, Zhang Y, Wu Q (2021) Traffic flow prediction over muti-sensor data correlation with graph convolution network. Neurocomputing 427:50–63. https://doi.org/10.1016/j.neucom.2020.11.032
46. Taulli T (2021) Implementing AI systems: transform your business in 6 steps. Apress Media LLC, New York. https://doi.org/10.1007/978-1-4842-6385-3
47. Wognum N, Bil C, Elgh F, Peruzzini M, Stjepandić J, Verhagen WJC (2019) Transdisciplinary systems engineering: implications, challenges and research agenda. Int J Agile Syst Manage 12(1):58–89. https://doi.org/10.1504/IJASM.2019.098728.
48. Kashyap R, Senthil Kumar AV (2020) Challenges and applications for implementing machine learning computer vision. IGI Global, Singapore. https://doi.org/10.4018/978-1-7998-0182-5
49. Wang KJ, Rizqi DA, Hong-Phuc Nguyen HP (2020) Skill transfer support model based on deep learning. J Intell Manuf. https://doi.org/10.1007/s10845-020-01606-w
50. Lin, Q, Zhang Y, Yang S, Ma S, Zhang T, Xiao Q (2020) A self-learning and self-optimizing framework for the fault diagnosis knowledge base in a workshop. Robot Comput Integr Manuf 65:101975. https://doi.org/10.1016/j.rcim.2020.101975
51. Zaatari SE, Wang Y, Li W, Peng Y (2021) iTP-LfD: Improved task parametrised learning from demonstration for adaptive path generation of cobot. Robot Comput Integr Manuf 69:102109. https://doi.org/10.1016/j.rcim.2020.102109
52. Kana S, Tee K-P, Campolo D (2021) Human-Robot co-manipulation during surface tooling: a general framework based on impedance control, haptic rendering and discrete geometry. Robot Comput Integr Manuf 67:102033. https://doi.org/10.1016/j.rcim.2020.102033
53. Tan W, Huang L, Kataev MY, Sun Y, Zhao L, Zhu H, Guo K (March 2021) Xie N (2021) Method towards reconstructing collaborative business processes with cloud services using evolutionary deep Q-learning. J Indus Inform Integr 21:100189. https://doi.org/10.1016/j.jii.2020.100189

Chapter 6
Object Recognition Methods in a Built Environment

Josip Stjepandić and Markus Sommer

6.1 Introduction

As a longstanding, fundamental challenge in field of computer vision, object recognition has been an emerging topic in the past decade. The goal of object recognition is to determine whether or not there are any instances of objects from the given categories (such as machines, cars, humans, bicycles, animals, or plants) in given image, video, or point cloud and, if present, to return the spatial location, orientation, and extent of each object instance (e.g. via a bounding box). As the cornerstone of image understanding and computer vision, object recognition forms the basis for solving more complex or high-level vision tasks such as segmentation, scene understanding, object tracking, image captioning, event detection, and activity recognition as well as subsequent process simulation and animation. Object recognition discovers a wide range of applications in many areas of artificial intelligence and information technologies, including robot vision, autonomous driving, security, quality assurance, consumer electronics, human computer interaction, content-based image retrieval, intelligent video surveillance, and augmented reality [1] which have relevance for the Digital Twin.

With current digital trends and developments, a close interconnection between the digital and the real world is necessary. Digital offering gets an even higher weight and the market share of products and services which are not yet affected by digital

J. Stjepandić (✉)
PROSTEP AG, Dolivostr. 11, 64293 Darmstadt, Germany
e-mail: josip.stjepandic@prostep.com

M. Sommer
isb innovative software businesses GmbH, Otto-Lilienthal-Strasse 2, 88046 Friedrichshafen, Germany
e-mail: sommer@isb-fn.de

J. Stjepandić et al. (eds.), *DigiTwin: An Approach for Production Process Optimization in a Built Environment*, Springer Series in Advanced Manufacturing,
https://doi.org/10.1007/978-3-030-77539-1_6

transformation is dramatically decreasing [2]. This also applies if the built environment exists that has not yet or just partially been digitalized because it cannot be included in a digital offering. One of the most frequent questions in the field of digital transformation is how to handle the physical objects which do not exist as an appropriate digital model. Such objects include complex long-living spatial objects like machines and industrial equipment, in particular which are produced once and used in operation and maintenance in a long period [3]. In the era of the Digital Twin, which simultaneously comprises the description of the real and virtual product, the requirement to make a virtual representation of any existing physical object is a pre-requisite for further procedure [2].

In the past few years, a wide popularization was achieved in 3D distance measurement equipment and low-cost 3D data acquisition devices (mobile phones, tablets, Google Glass, Microsoft Hololens) based on digital platforms, which easily provide dynamic 3D data in a satisfactory quality [4, 5]. Significant research and development have been accomplished in contactless 3D measurement, 3D object reconstruction, 3D object localization, and 3D modeling based on the outcome of these devices [6]. A great challenge still remains in how to transform semantic unstructured or low-level digital data (e.g. a high-quality image, video, or point cloud) of an (industrial) built environment in a compact, process-oriented digital representation which can be seamlessly used for generation of Digital Twin [7–9]. This transformation of acquired data lies in the focus of the research and development presented here.

In a rough industrial environment, which is accompanied with several environmental, potentially negative impacts (dirt, dust, smoke, darkness, vibration), object recognition is much more challenging than the simple detection of an obstacle in the space: each object must be put in the exact position and get the right orientation in space, and identification attributes [10]. In practical measure, object recognition should replace a significant extent of the manual task (manual remastering) which is tedious, time-consuming, inaccurate, and a source of mistakes (Fig. 6.1) [11]. Hence it is also called reverse engineering because it creates the missing CAD model of the physical object [7].

While the Digital Twin is undertaken the frequent change of the geometrical shape of an object, a high semantical and geometrical accuracy in the process of the data acquisition is required to describe the object extension (shape) in the space [12]. Subsequently, the high-quality CAD data of all geometrical objects in all stages of planning process are the pre-requisite for seamless downstream processes [13]. Therefore, the quality of data generated during the recognition process should be comparable with the quality of data generated by standard, methodical, manual engineers' work.

Object recognition basically comprises multiple objects of different categories in a scene (e.g. factory hall). The generic process of the object recognition is depicted in Fig. 6.2. At first, an object must be acquired by an observation which can be done by devices through various physical principles (photo, video, distance measurement) as described in Chap. 4. Recognition methods with all basic principles will be described further. In the next step, the features which can be identified in the scene must be classified against a patterns database. The recognized features are integrated in the

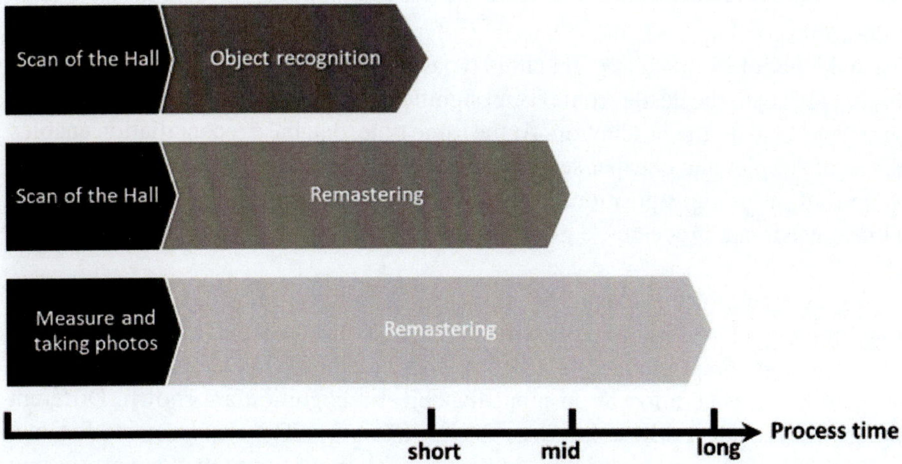

Fig. 6.1 Approaches for shape acquisition of physical objects

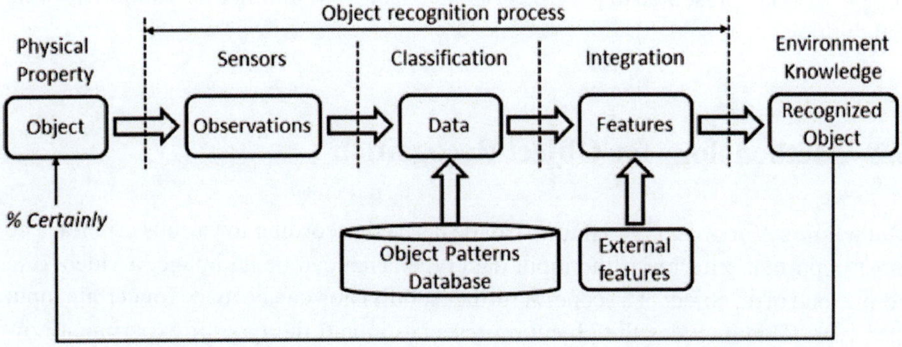

Fig. 6.2 Object recognition process [8]

next step to objects by using external features. By using environment knowledge, the result of this process is the recognized object, based on a recognition accuracy.

Objects in a factory have some distinct external characteristics: some of them are enclosed by a cover (e.g. machines), the others (e.g. pipelines) discover a distinct dimension. Even if objects in the factory often touch each other, there is rarely an overlap, but the visual occlusion is very common. From a geometrical point of view, they consist almost exclusively of standard geometries with many prismatic parts. This opens up the opportunity to combine the steps of classification and integration in Fig. 6.2 by searching for and recognizing the entire objects rather than the individual features which would be collected to an object further. This assumes that the external shape of an object is distinctive enough and the object pattern database large enough to ensure that the object can be always recognized.

Embedded in a working environment, the practical procedure can appear as follows: instead of identifying the detail feature of each object, the identification of the corresponding CAD models from a reference library and the transfer of geometry and other object data (e.g. machine types) as modular objects directly from the library [11, 12] to the design context can significantly reduce the scan times for a first rough "prescan" of the production. At the same time, database reconciliation enables the use of simpler and cheaper search methods [14]. After a specific object has been recognized in the right position, its corresponding native CAD model can be used in the downstream processes, e.g. material flow simulation to generate Digital Twin [15].

The remainder of this chapter is structured as follows: in Sect. 6.2, the methodology for object recognition of semantic model is presented by distinguishing between image, point cloud, and video-oriented methods. In Sect. 6.3, approaches for point cloud generation as an input for object recognition are shown. Different approaches for the test base are described in Sect. 6.4. The requirements for data acquisition accuracy are highlighted in Sect. 6.5. Methods for point cloud processing are described in Sect. 6.6, followed by the comparison of recognition methods in Sect. 6.7 with the related application of methods in Sect. 6.8. Discussion and future perspectives are presented in Sect. 6.9. Finally, Sect. 6.10 summarizes the conclusions and outlook.

6.2 Methodology for Object Recognition

Methodology for object recognition can be derived according to various criteria. The most important criterion is the input data type. That can be an image, a video, or a point cloud of an object or a scene. A further subdivision can be made for certain input data type. This includes the object representation and the type of experiment how the sample results are achieved. A large test base is necessary for both the detection of the progress in development of an object recognition method and comparison of results achieved by different object recognition methods.

In the case presented here, the input data type is decisive for the structure of the entire process chain from the initial data acquisition until generation of Digital Twin. That is why it is covered in detail as the main criterion. The solution alternatives will also be drawn.

6.2.1 Challenges in Object Recognition

The classification of the challenges in object recognition is illustrated in Fig. 6.3. The ideal goal of generic object recognition is to develop general-purpose object recognition algorithms achieving two competing goals: high quality and accuracy as well as high efficiency. High quality recognition has to accurately localize and recognize objects in images, point clouds, or video frames, such that the large variety of object categories in the real world can be distinguished (i.e., high distinctiveness), and that

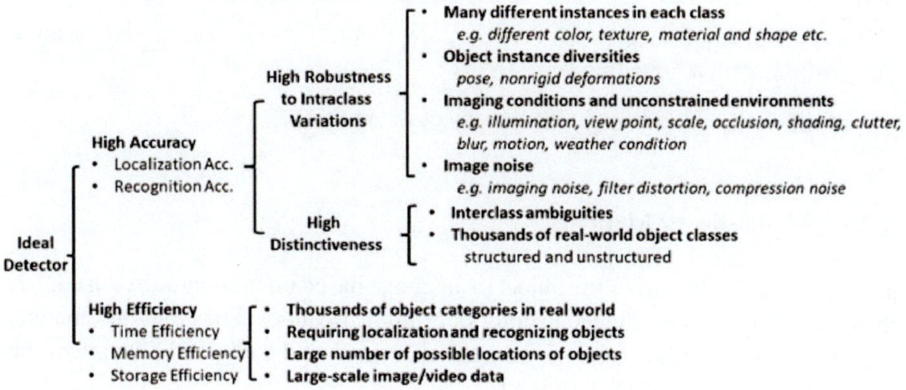

Fig. 6.3 Summary of challenges in generic object detection [1]

object instances from the same category, subject to intraclass appearance variations, can be localized and recognized (i.e., high robustness). High efficiency requires the entire detection task to run at a sufficiently high frame rate with acceptable memory and storage usage [1].

6.2.2 *Image-Oriented Methods*

We, as human beings, recognize objects through the recognition process in the brain. The brain stores the patterns of individual objects. In humans the process happens emotionally, an artificial system must extract the objects from the individual pixels of the image.

In the case of a colour camera, each pixel represents a shot of the light the camera receives from the direction of the object. The pixel has information about colour and brightness for the area of the pixel. The colour and brightness depend on the angle and type of illumination, as well as the reflection of the object. Strong changes of the pixel values can be an indication for a transition from one object to another, gradual changes indicate a change of the surface orientation (light cone).

For object recognition, a data collection of the objects to be recognized is required. An image can be acquired with different systems such as colour, infrared cameras, radar, or X-ray devices, etc. In the data collection the individual objects on the image are annotated [16].

In object detection, the individual objects are detected with bounding boxes. The recognized objects are to be assigned to types or classes of objects. An overlapping of the individual objects or limitation frames is possible at any time [17]. The object recognition is divided into two areas:

– Classification: With the classification the recognized object is assigned to a type or class.

- Object localization: Search for the recognized object in an image and determine its position with a bounding box.

The following methods are used for object recognition.

6.2.2.1 Model-Based Methods

The edge detection identifies the object in an image based on the contours. An edge is considered to be a strong change in color or brightness values. The larger the change, the more intense the edge. By analyzing the alignment of the edges, the length of the object in the image can be determined.

With active contours approach, an initial closed contour with defined properties is searched in the image and the calculation process is started. The specifications are optimized from the properties of the contour and from other properties of the image (features). Known representatives of these active contours are snakes, finite elements, and spring-mass models [18].

6.2.2.2 Appearance-Based Methods

For histograms, the distribution of intensity values of an object is calculated. The course of the histogram is described by the respective object like a template. For the recognition of the object the individual histograms are calculated and compared with the known templates (Fig. 6.4). If the course of the histogram matches the course of the template, the object has been detected [19].

The color DH consists of two parts the color histogram and geometric information. The geometric information is generated by pixel color pairs with a certain distance in the image space. The two pieces of information classify and describe the position of the object in the image [19].

6.2.3 Point Cloud-Oriented Methods

When recognizing 3D objects, a further feature of the image-based methods is the depth information for the recognition of objects. In the following some methods are described as examples [20, 21].

6.2.3.1 Feature Based Methods

Feature-based methods are rest on the evaluation of the representations and are the most common type of evaluation. These features describe 3D objects, i.e. several features (descriptors and key point detectors, feature vector constructions) describe

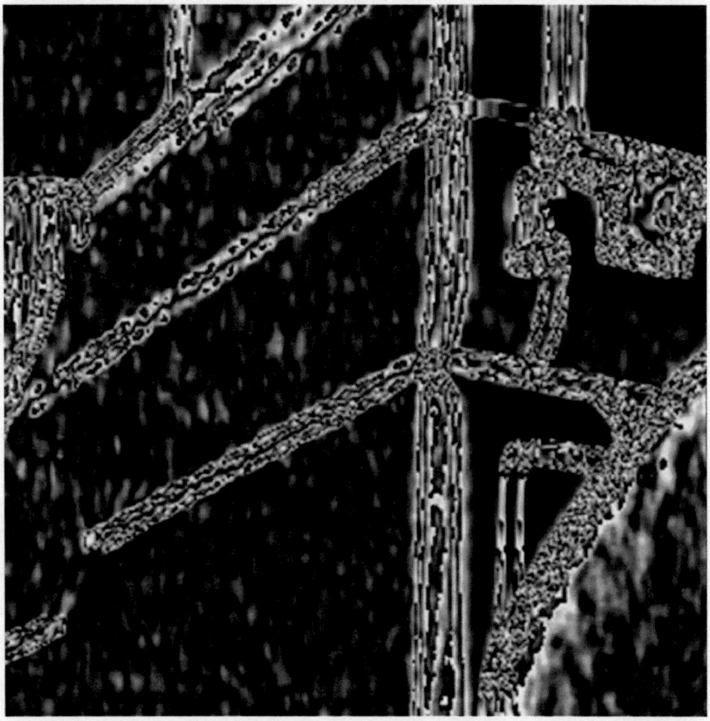

Fig. 6.4 Calculation of an image with convolution algorithms

the appearance of the object in space with its geometric features and shapes. There are two approaches to consider features in a global or local context. In the global context the combination of features in space must be unique, which means a high demand on the calculation of features. However, there are cases in object recognition, e.g. several similar objects, where the similar objects must be described globally with unique features. The other way is to calculate features locally in a clearly defined area in space. The features can then be calculated more accurately and better. A better recognition is thereby ensured. To optimize the object recognition, both approaches (global and local) are combined. First a coarse granular recognition of the 3D objects in categories (e.g., machine, conveyor line) is performed. With the classification into the category, local features for the recognized category are explicitly used for this area. The result of the analysis of the local features describes the 3D object completely.

6.2.3.2 View-Based Methods

The view-based methods capture the entire object in its appearance. An appearance would be e.g. the outer edge of a chair in a room. The outer edge of the chair can vary according to the angle of view of the object or even change completely, e.g., when the chair is viewed from below, the seat and the four legs of the chair can be evaluated. I.e., if the 3D object is viewed from an unfortunate angle, the object cannot be recognized if the recognition algorithm does not know the pattern of the progression. In this case there are two methods to improve the recognition rate. Analyzing the object with the best views and assigning these patterns to the object for recognition. Another method is to analyze the object from several views (15° each) and assign several patterns to the object. In the case of the chair, the method involves extracting the necessary properties. The chair is still a simple case. If the 3D objects become more complex, combined methods between feature-based and view-based methods are used very quickly to increase the efficiency and recognition rate.

6.2.3.3 Graphs

The graph is an abstract structure in graph theory, in which a set of properties and their connections to each other are described. The properties are called nodes (also corners) of the graph. The connections between properties are called edges. The edges have two characteristics: directional and non-directional. In the representation of graphs, the nodes are shown as points and the edges as lines [22].

The different properties of the point cloud are extracted, classified and arranged in the graph (Fig. 6.5). The graph describes the object to be recognized by the individual dependencies of the properties on each other. With this approach also 3D objects can be recognized, which correspond to less than 90 percent of the known object.

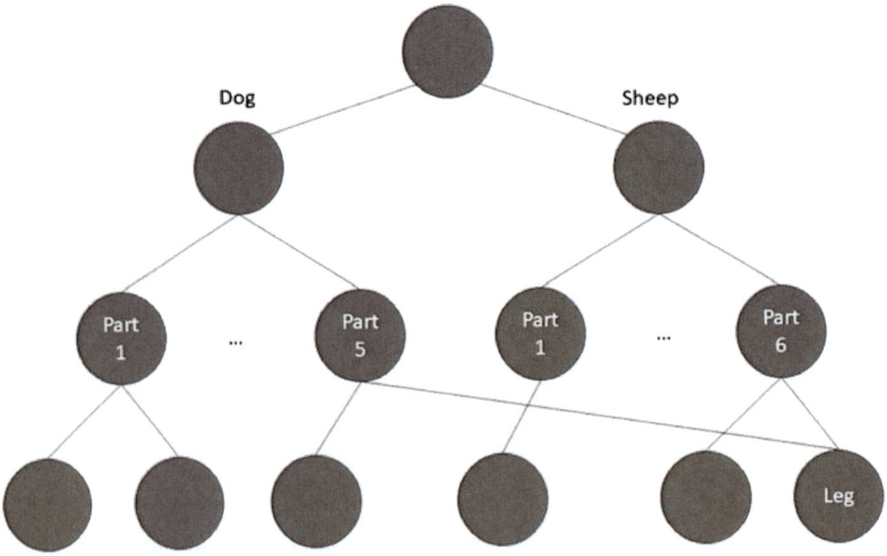

Fig. 6.5 Graph display to distinguish between Dog and Sheep with the same properties for the Leg

6.2.4 *Video-Oriented Methods*

A video is a sequence of images and describes the object to be recognized. With this description of the object, the individual images cannot be analyzed individually. The temporal component is added to the analysis. In addition, a video can be viewed as a sequence of individual snapshots of the environment. For the analysis of the video two areas are used. The known area from image-based object recognition and the analysis of the change of the object over the single images. An object does not only have to be filmed by one camera, several different cameras can be used at different angles. The image sequences of the individual cameras have to be synchronized with each other in order to merge the angles of view later [23].

6.2.4.1 Globalshutter Versus Rollingshutter

CMOS sensors with global shutter enable sharp images for image processing, since each pixel is exposed simultaneously. In contrast to rolling shutter sensors, which perform exposure sequentially by line, column or pixel, not all pixels are exposed at the same time. The exposure time for an image extends over a frame period (one per frame rate) until all pixels on the sensor were light sensitive. If an object or the recording camera moves very fast, distortions are visible on the images. Especially if the exposure time (e.g., in dark environments) is very high, the distortion becomes more visible in the image. Rolling shutter cameras are not suitable for an efficient evaluation of the images and deteriorate the results for further analysis of the image right from the start. In the end, several images merge into one image, which makes it difficult to identify the objects clearly over time. Subsequent algorithms have to take this fact into account and react tolerantly to inconsistencies at this point. Especially in the field of feature extraction, a consistent acquisition of the individual images is the basis for a very good feature recognition with a Global shutter sensor [24].

6.2.4.2 Feature Tracking

Feature recognition is used to extract individual features from the images. The features can be extracted using well-known extraction algorithms such as SIFT, ORB, etc. All these algorithms calculate a point with a corresponding descriptor. The descriptor describes the environment of the point. Both properties have to be calculated in a way that the same point respectively descriptor gives the same value from other angles. With this property points respectively descriptors can be tracked over single film sequences. However, it is not important to find many factors. It is important to recognize the features clearly over many images of a video sequence.

6.2.4.3 Pose Graph

The pose graph is the movement of the camera over the individual images in space. The coordinate system starts at the point *0,0,0* with the first image of the video system. By tracking at least ten or more similar features (point or descriptor) over the individual images, the position of the camera in space and the image direction is calculated. The pose graph also describes the relationships of the individual images to each other. The more points can be used for the calculation, the more reliable the position in space can be determined. A thin cloud of points results from the pose graph and the recognized feature. To improve the pose graph, maps are used to detect recurring images. If an image is recognized, the already calculated pose graph is recalculated in the context of the recognized image. This is then called a closed loop [25].

6.2.4.4 Depth Maps

The individual depth maps are initially calculated from the pose graph, i.e., every connection between two nodes in the graph creates a depth map. The more depth maps we generate from one viewing direction to an object, the more we can optimize the depth maps to each other. The depth maps are combined with the images from the video film [25].

6.2.4.5 Object Recognition

The steps feature extraction, (Pose Graph or Depth Map leading to a 3D Image) generate a lot of information about the object (Fig. 6.6). The more information we have about the object to be recognized, the more possibilities we have to clearly recognize the object in space. Especially the image and video-based methods show advantages in the evaluation of the images. By adding depth information to the video-based methods, the object can be recognized even more selectively [25].

6.3 Approaches for Point Cloud Generation

In general, the input for 3D reconstruction techniques is the output of data collection devices as described in Chap. 4. If it is intended to conduct the object recognition based on the point cloud of the corresponding objects, as for generation of Digital Twin presented here, it is necessary to prepare the input data generated by different types of cameras and scanners.

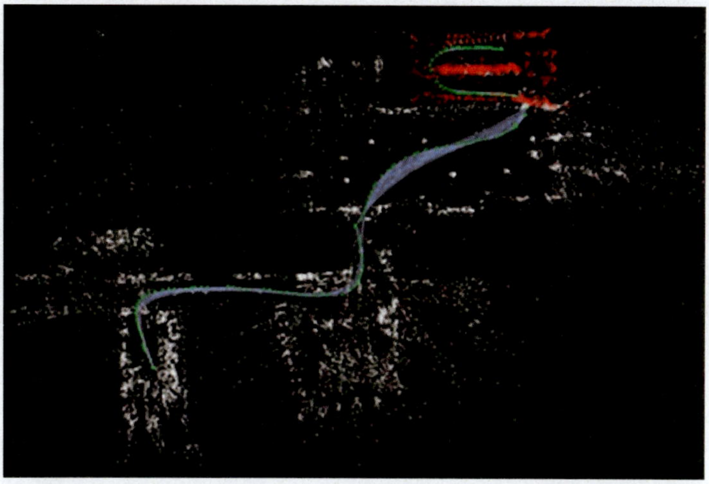

Fig. 6.6 Pose graphs with point cloud of an indoor scene

In this section, the processes of generating point clouds from monocular images, stereo images, and video frames as well as methods used in these processes are presented. Since the outputs of laser scanners are point clouds (beside auxiliary images), a generation of point clouds is obsolete [10]. A generic process flow for different types of input data is illustrated in Fig. 6.7 on the left-hand side. Distinction

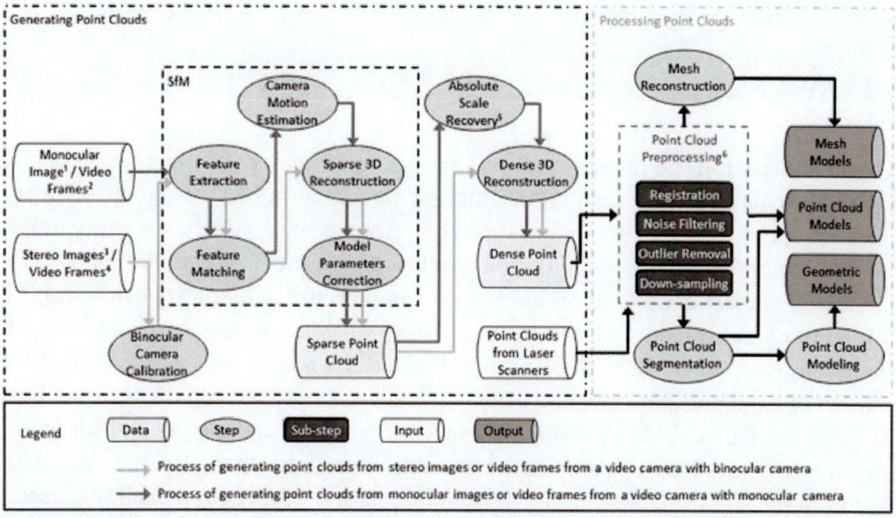

Fig. 6.7 The framework of 3D reconstruction techniques in engineering [10]. *Note* 1-generated from monocular cameras; 2-generated from a video camera with monocular camera; 3-generated from binocular cameras; 4-generated from a video camera with binocular cameras; 5-the step is not mandatory in some cases; 6-sub-steps here are not mandatory, and which sub-steps are needed depend on the application requirements and the quality of the point clouds

is made between monocular and stereo images respectively video frames in the following way:

1. Generated from monocular camera;
2. Generated from a video camera with monocular camera;
3. Generated from binocular camera;
4. Generated from a video camera with binocular camera.

Processing of point clouds is described in Sect. 6.5.

The block named Structure for Motion (SfM) comprises the five mandatory steps for all types of cameras: feature extraction, feature matching, camera motion estimation, sparse 3D reconstruction, and model parameters correction which yield a sparse point cloud as the result. In case of a monocular camera, an absolute scale recovery may become necessary (note 5 in Fig. 6.7). In case of binocular camera, an initial camera calibration is necessary. While sparse cloud is used to detect which image pairs are overlapping (by sufficient number of common valid tie points), the dense point cloud comprises the depth maps for such overlapping pairs which were merged together.

Subsequently, a dense 3D reconstruction follows with a dense point cloud as the result which is input for the point cloud processing, e.g. object recognition (further description in Chap. 8). The aim of this step is to recover the details of the scene, by using the images of a certain scene, the intrinsic and extrinsic parameters of cameras for each image after model parameters correction and the sparse point cloud obtained from the previous step to generate a dense point cloud.

6.4 Test Base

A critical aspect in the procedure for object recognition is the proper test base. To train a deep 3D shape representation with a significant intra-class variance requires a large collection of 3D shapes. Since this requirement usually appears in every object recognition project, there are a limited number of datasets that can be used to train and test 3D object classification methods for indoor and outdoor applications [26]. No international standard is known on this area at this time. Many CAD datasets, created in the past decade, were limited both in the variety of categories and the number of examples per category. These data sets can be used to train the classifier.

However, the publicly available data sets are synthetic and too general for industrial applications. Hence, such a test base will never fully fit the specific problem. Accordingly, appropriate training with suitable training data is essential. It is necessary to build-up an own object library with additional objects and train the object recognition framework with them. This is the only way to achieve high accuracy and low overfitting at the same time [27]. Table 6.1 shows a brief list of frequently used 3D data sets: ModelNet in two splits, ShapeNet and Sydney Urban Objects. Their focus lies on the object recognition of single objects.

Table 6.1 Popular 3D data sets that focus on studying individual objects along with their number of samples and classes [26–28]

Data set	#Samples	Categories	Classes
ModelNet10	4,899	10	10
ModelNet40	151,128	660	40
ShapeNet core	51,300	–	55
Sydney urban objects	631	–	14

Here the data set ModelNet [26] is one of the most used for the research of 3D object recognition. It contains two splits, ModelNet10 and ModelNet40, which are used to classify household items such as chairs and desks as well as industrial objects like airplane and car [28]. ModelNet is used today for the basic assessment of new recognition approaches.

Princeton ModelNet library provides researchers with a comprehensive clean collection of 3D CAD models for objects. To compose ModelNet, 3D CAD models from 3D Warehouse, and Yobi3D search engine indexing 261 CAD model websites were downloaded. Common object categories were queried from the SUN database with more than twenty object instances per category, removing those with too few search results, resulting in a total of 660 categories. Once a vocabulary for objects was established, 3D CAD models belonging to each object category were collected using online search engines by querying for each object category term. Then, human workers were hired to manually decide whether each CAD model belongs to the specified categories, using an in-house designed tool with quality control. The authors then manually checked each 3D model and removed irrelevant objects from each CAD model (e.g, floor, thumbnail image, person standing next to the object, etc.) so that each mesh model contains only one object belonging to the labeled category. Unrealistic (overly simplified models or those only containing images of the object) and duplicate models were also discarded. Compared to a previous benchmark, which consists of 6,670 models in 161 categories, the new dataset is 22 times larger containing 151,128 3D CAD models belonging to 660 unique object categories. Finally, ten popular object categories were chosen, to obtain a very clean dataset and manually deleted the models that did not belong to these categories. Furthermore, the orientation of the CAD models for this ten-class subset were manually aligned as well. The ten-class subset and the full dataset are available for download [26, 28].

ShapeNet is a richly-annotated, large-scale repository of shapes represented by 3D CAD models of objects. It contains 3D models from a multitude of semantic categories and organizes them under the Word-Net taxonomy. It is a collection of datasets providing many semantic annotations for each 3D model such as consistent rigid alignments, parts and bilateral symmetry planes, physical sizes, keywords, as well as other planned annotations. Annotations are made available through a public web-based interface to enable data visualization of object attributes, promote data-driven geometric analysis, and provide a large-scale quantitative benchmark for research in computer graphics and vision. ShapeNetCore is a subset of the full ShapeNet dataset with single clean 3D models and manually verified category and alignment annotations. It covers 55 common object categories with about 51,300 unique 3D models. The twelve object categories of PASCAL 3D+, a popular computer vision 3D benchmark dataset, are all covered by ShapeNetCore [29].

The fourth dataset, Sydney Urban Objects, is a small point cloud dataset. The point clouds were measured by a LIDAR sensor that examines an urban environment. There are 631 individual scans of objects across classes of vehicles, pedestrians, signs, and trees. In contrast to the other datasets mentioned above, the data is taken from reality and is thus filled with noise. It aims to provide non-ideal sensing conditions that are representative of practical urban sensing systems, with a large variability in viewpoint and occlusion. This, however, comes closer to the real application case and is therefore preferred [30].

6.5 Impact of Data Acquisition Accuracy

Prior to each process and method development, a detailed analysis of the requirements for accuracy of processed data is necessary. As described in Chap. 4, the selection of the dedicated scanning device and procedure determines the entire process of the object recognition. It has a significant impact to the result, quality, accuracy of recognized objects, and process time. It is, however, recognized that the non-measured or missed points, features, or building parts remain an open problem, e.g. in case of occlusion or areas in shadow (for example on the roof).

Although various parameters have been proposed, the two most common parameters for point cloud data quality are accuracy and spatial resolution which determines the density of the point cloud [31]. The parameter accuracy measures the ranging precision of each single point. This is mainly determined by the hardware performance of the equipment as well as the data acquisition parameters [7]. No benchmark for accuracy of scanning devices (scanner, camera) is known at this time.

The other parameter, spatial resolution, measures the spatial density of point cloud data. It is mainly determined by the data acquisition parameters. In some cases, coverage is mentioned as another parameter of point cloud data quality [7]. The determination of parameters for point cloud acquisition is necessary to fulfill the required data quality and to carry out the acquisition in a time-effective manner [7].

Each factory and each building have their own history and life cycle. Most commonly, the status "as-built" is poorly documented and must be thoroughly captured with a sufficient accuracy in case of change, reorganization and retrofit [10]. Objects in the factory have various extensions and are as components built with different accuracy requirements. Nevertheless, the object recognition in the approach presented here is conducted with a constant accuracy for all objects in a scene [11].

The point cloud quality was addressed in three ways, by defining levels of quality parameters, by evaluating the quality parameters of a point cloud, and by defining parameters of a scanning plan in order to achieve a desired level of quality. The U.S. General Services Administration (GSA) published guidelines for the solicitation of 3D imaging services and evaluation criteria to ensure that the specified requirements for the deliverables are met (Table 6.2) [32].

Table 6.2 Data quality definitions introduced by GSA [32]

Level of detail	Level of accuracy (tolerance) [mm]	Minimum artefact size [mm]
Level 4	±3	13 × 13
Level 3	±6	13 × 13
Level 2	±13	25 × 25
Level 1	±51	152 × 152

Table 6.3 The accuracy of the outputs of 3D reconstruction techniques [10]

Inputs	Outputs	Objects	Dimension differences (mm)
Monocular images	A dense point cloud	Indoor space	1.4
		Outdoor space	2.7
		A school building	107
		An elecrical substation	7.1
		Cracks	0.34
	Dense point clouds	Buildings	20
	A mesh model	A column	7
Video frames	A sparse point cloud	A construction site	4.7
Point clouds from laser scanners	A point cloud	Roads	6
		Pipelines of MEP systems	4
	A mesh model	Walls	43.6
	A geometric model	An interchange bridge	190
		Road junction	5
		Pipelines of MEP systems	3.8

Four arbitrary levels of detail (LOD) have been defined, ranging from minimum artefact size of 13 × 13 mm to 152 × 152 mm, and the level of accuracy (LOA) from 3 to 51 mm (Table 6.2). The LOD correspond with areas of interest, LOD 1 with the total project area, LOD 2 with the building, LOD 3 with a floor, and LOD 4 with a room or artefact [32]. Therefore, our focus lies on artefacts with level of detail four to two.

Ma and Liu have conducted an exhaustive survey of results published in journals [10]. The dimension differences of the outputs of 3D rebuild techniques range from a few millimeters to a few centimeters, as shown in Table 6.3, which is acceptable for the cases where their application requirements for accuracy are not high. Generally, point clouds from laser scanners have high accuracy, and they are taken as ground truth to determine the accuracy of point clouds generated from monocular images, stereo images, and video frames [10]. The values determined lie within the limits in Table 6.3.

In the current state, the point cloud is created with a high manual effort. The expert knowledge and experience of the engineers is also needed here. On the one hand, the collection of insufficient data makes a model useless for the intended application. On the other hand, the collection of too much data requires more time and effort and leads to redundant data. Accordingly, it is important to find a suitable degree and to determine the required quality of the point cloud for each specific application (s. Chap. 2). With this, the acquisition of point cloud data can be facilitated [7]. By using a modern terrestrial scanner, a single scan with subsequent automatic registration takes approximately twelve minutes. For a hall with 1.000 m^2, up to 25 scans from different positions are necessary.

It is worth to note that depending on the object recognition framework the 3D meshes are often converted into voxel s because they can be convolved like any other tensor. Estimating the voxel orientation in addition to the classification performed can improve the overall results [33].

6.6 Methods for Point Cloud Processing

The rough point clouds of scenes in a factory, gained either by scans or from images and videos with subsequent dense point generation, cannot be used directly for the object recognition objects for several reasons. Such a point cloud contains false or undesired information, e.g. surroundings or reflections by large glass surfaces. Therefore, the point cloud preprocessing (as marked by note 6 in Fig. 6.7, on the right-hand side) is necessary to provide a point cloud model for seamless further processing, e.g. object recognition.

Four transforming steps are needed to make the point clouds of a scene meet the application requirements: point cloud preprocessing, mesh reconstruction, point cloud segmentation and point cloud modeling. The algorithms and methods used in each step and sub-step are shown in Table 6.4.

6.6.1 *Point Cloud Preprocessing*

The aims of the step are to prepare the point cloud gained by a device (scan, camera) for further processing in several transformation and translation steps by provision of point clouds with high quality for downstream operation. Generally, the step contains four sub-steps, i.e. registration, noise filtering, outlier removal and down-sampling. It deserves to explain that which of the four sub-steps are needed depends on the

Table 6.4 Algorithms and methods used in for processing point clouds [10]

Steps	Sub-steps	Algorithms and methods
Point cloud preprocessing	Registration	Iterative Closest Point (ICP)
	Noise filtering	Removing points manually
	Outlier removal	Removing points manually, RANSAC
	Down-sampling	Point spacing strategy
Mesh reconstruction	–	Poisson surface reconstruction (PSR)
Point cloud segmentation	–	Region growth
		K-means clustering
		Voxel-based algorithm
		Hough transform
		RANSAC
Point cloud modeling	–	Obtaining the dimensions of objects

application requirements and the quality of the point clouds. For example, if the application requirement is a combined point cloud and the existing point clouds are separate, then registration is required which merges two point clouds into a consistent point model by a spatial transformation [10].

Typically, the algorithm used for registration is the iterative closes point (ICP) algorithm, which estimates the rigid transformation between two point clouds iteratively by minimizing the distance between matched points [10]. Modern scanner provides an automatic registration function. In some cases (e.g. distance between two scanner positions is too large or significant occlusion) the registration must be done manually by a point cloud editor.

Point clouds obtained from Sect. 6.3 incorporate not only the points from the objects of interest, but also the points from surroundings (also known as noise points) and outliers as shown in Fig. 6.8. Noisy data makes impossible applying key point detection and feature extraction techniques. The non-desired reflections of objects in an office happened on large glass surfaces (Fig. 6.8 left-hand and bottom side). The process of removing noise points and outliers are noise filtering and outlier removing respectively. Generally, the algorithms and methods used for noise filtering can also be used for outlier removing and vice versa, because noise points and outliers are all unwanted points and they are not from the objects of interest [10]. The algorithms and methods used for noise filtering and outlier removing include removing points from surroundings manually and by using random sample consensus (RANSAC).

In addition, point cloud registration makes overlapped regions denser, which will make the data handling difficult and, thus, reduce the efficiency of subsequent processing. By down-sampling such problems are resolved. The algorithm generally

Fig. 6.8 The rough output of the scan procedure in an office [11]

used for down-sampling is point spacing strategy, which can reduce points in dense regions. In the method, as an algorithm, a point cloud is arranged into voxels (3D grid cells) with equal sizes and points in every single 3D grid cell containing at least one point are reduced until requirements met [10]. After this step, the data volume of a data set is reduced by factor five to ten respectively which yields much easier handling and shorter lead time without losing accuracy.

6.6.2 *Mesh Reconstruction*

The aim of the step is to generate a mesh model of the object of interest by using the point cloud obtained from previous step in cases when the object recognition does not directly process the point could. Typically, the algorithm used for mesh reconstruction is Poisson surface reconstruction (PSR), which reconstruct a watertight, triangulated approximation to the surface of the object according to its point cloud. The output surface mesh is generated by extracting an iso-surface of this function. Starting with unorganized point cloud data, unexpected triangles, such as holes and slits, may be generated during mesh surface reconstruction [10]. To repair the mesh holes that can easily be generated by the Poisson surface reconstruction algorithm, an iterative postprocessing algorithm for patch generation is used, which also reduces the amount of reconstructed mesh data.

Mesh reconstruction is beneficial further. It is necessary for some application requirements due to two reasons:

(1) Compared to dense point clouds, mesh models obtained from mesh reconstruction are the better choice for visualization of the objects of interest;

(2) Mesh models can be used for subsequent applications, such as crack detection or voxelization [10].

6.6.3 Point Cloud Segmentation

The aim of the step is to segment a point cloud and obtain the points of the objects of interest. Point clouds are classified into multiple homogeneous areas, the points in the same region will have the same properties. The segmentation is challenging because of high redundancy, uneven sampling density, and lack of explicit structure of point cloud data. Although algorithms used in publications differ slightly from each other, the algorithms used for point cloud segmentation can be classified into five categories based on their design mechanisms (Fig. 6.9): edge-based, region-based, attribute-based, model-based and graph-based methods.

However, in general, there are two basic approaches. The first approach uses purely mathematical model and geometric reasoning techniques such as region growing or model fitting, in combination with robust estimators to fit linear and nonlinear models to point cloud data. This approach allows fast running time, achieves good results in simple scenario. The limitations of this approach are it is difficult to choose the size of model when fitting objects, sensitive with noise, and not working well in complex scenes. The second approach extracts 3D features from point cloud data using feature descriptor, and uses machine learning techniques to learn different classes of object types, and then uses the resultant model to classify acquired data [34].

In complex scenes, the machine learning techniques will outperform techniques purely based on geometric reasoning and, thus, should be preferred. The reason is due to noise, uneven density, occlusions in point cloud data, it is very difficult to find and fit complicated geometric primitives to objects. Although machine learning techniques give better results, they are usually slow and rely on the result of feature extraction process [34]. An important information that has been neglected in the development of 3D point cloud segmentation algorithms is integration of contextual information, e.g. assignment to dedicated object groups: building, machines, pipelines, etc..

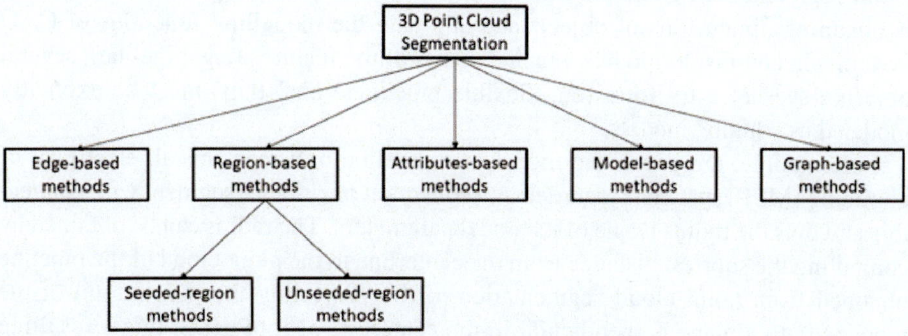

Fig. 6.9 Taxonomy of 3D point cloud segmentation methods [34]

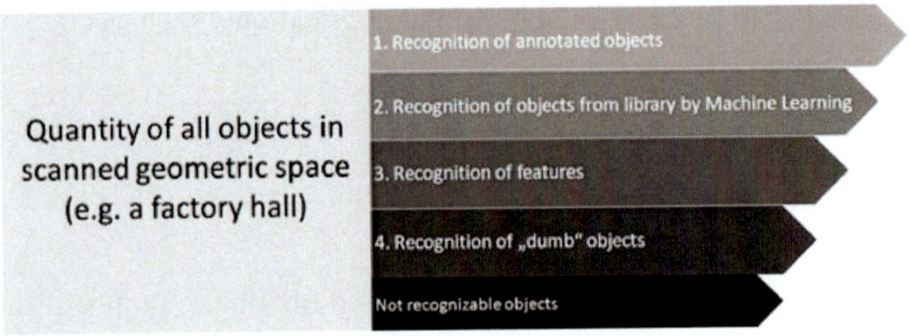

Fig. 6.10 Object recognition in detail [11]

6.6.4 Point Cloud Modeling

The aim of the step is to generate a geometric model of the object of interest by using the point cloud segment obtained from the previous step. Basically, there are four ways to create a model (Fig. 6.10):

(1) Identification of an annotated model (or object) and its position,
(2) recognition of a non-annotated model (or object) from a model library and its position,
(3) recognition of the singular features of a model and the composition of the model itself by recognition of several features in the right position, and
(4) the recognition of "dumb" objects by manual remastering of a model using a point cloud segment.

It is worth to note that some portions of point cloud remain not recognizable.

For identification of annotated model (or object) a complex procedure based on scanning is not necessary. It is applicable for new, well-documented objects only. For recognition of a non-annotated object an appropriate method as described in Sect. 6.3 with accompanied model library is necessary. This will be described in the following sections.

In order to generate a geometric model (variants 3 and 4 in Fig. 6.10) of the object of interest, it is essential to obtain the model parameters based on the results of point cloud segmentation. Generally, the method used for obtaining model parameters is obtaining dimensions of objects and applying the modelling functions of CAD systems. In contrast to models which consists of the features (e.g. pipeline), several objects have not a fix form (e.g. flexible pipelines) and, thus, must be explicitly modeled as "dumb" models.

For example, the geometric model of a pipeline of a mechanical, electrical, or plumbing (MEP) system is regarded as a collection of cylinder segment with connectors and thus the radius is one of its model parameters. The radius can be obtained by computing the shortest distance from the centerline to the point cloud of the pipeline obtained from point cloud segmentation, or by projecting the point cloud of the pipeline onto a plane perpendicular to the centerline of it to generating a resulting circle and computing the radius of the circle [10]. With the model parameters, a geometric model of object of interest can be easily generated.

Although both the mesh models and the geometric models can be used for object visualization, the former can be used for representing crack based on its triangle mesh, while the latter cannot, because a crack can hardly be represented by using a geometric model. Furthermore, most CAD systems can use a point cloud as a passive (reference) geometry with a reduced functionality. However, such models can be used in the downstream processes because simulation modules require high-quality feature models.

6.7 Comparison of Recognition Methods

After in Sect. 6.2, the classification of methods for the recognition of a dedicated object in a section of a factory were presented, the focus in this section, however, is constricted to the recently developed Deep Learning (DL) techniques based on point clouds [24]. This trend is heavily supported by development of lidar devices, as a cheap, efficient, and rapid remote sensing technique. Deep learning, through an end-to-end model, directly performs a nonlinear transformation layer by layer on the original input from a lower level to a higher level. Hence, it can automatically extract the information contained in massive data, which greatly simplifies the learning process [35].

Basically, two different approaches can be used for objects recognition in point cloud data gathered with a scanner: point-based and voxel-based (after the transformation of point cloud into a regular structure). However, there are certain difficulties in using DL to process 3D point clouds [36] which must be considered:

- The point cloud is not in a regular format. Unlike an image, a point cloud is a set of points distributed in space. Hence, it is non-grid data. However, typical convolutional architectures can merely deal with highly regular data formats, and it is, therefore, hard to use Convolutional Neural Networks to process raw point clouds [35].
- One natural and naive representation of point clouds is set, which is unordered, irregular, and sparse, making its understanding difficult [37]. There are various orderings of points, indicating that there are many matrix representations of a particular point cloud. Therefore, adapting the changes of how the waypoints are arranged is a problem in point cloud processing [35].
- Three-dimensional point clouds normally contain only the spatial coordinates of points, lacking rich textures, colors, and other information.

To resolve these disadvantages, point clouds can be translated into voxels for further processing, although it yields high computational complexity. These operations are still based on sparse volumes, and it is challenging to deal with large-scale problems. However, this process chain pre-requisites several formal rules and quality of point cloud as described in previous Sects. 6.5 and 6.6. There are two downsides to voxels. First, 3D meshes and point clouds only encode the surface, so most of the voxels in a sample are zero valued. Second, the dimensionality of voxels increases cubically, which is why most applications keep the dimensionality small (e.g. 32 × 32 × 32), which eliminates most fine details [38]. Therefore, the voxelization reduces the differences between two similar objects with almost equal extension. For this reason, the distinction of two almost identical objects with small differences could be a serious issue and must be considered [39].

Based on an exhaustive literature survey, several recently developed object recognition approaches were selected and compared [27]. The most interesting approaches and their main characteristics are shown in Table 6.5 [39]. For the sake of completeness and comparison, two image-based approaches (3DShapeNets and ObjectNet3D) are also added at the bottom of this table [40]. Geometric shape descriptors (Principal Axle Descriptor [41] and Variational Autoencoder [42]) build a vector and search for similar objects in a library. G3DNet [33] and PointNet [43], as point-based methods, have a graph-based DL architecture that is used in 3D object classification and segmentation. Learning permutation invariance feature directly from raw 3D point clouds using deep neural network is a trend, which are effective and computationally efficient. In contrast, a 3D detection framework (VoxNet [44], VoxelNet [45], LightNet [46]) transforms the point cloud into voxels that each contain a small

Table 6.5 Main characteristics of different approaches for object recognition [39]

Approach	Input	Frame-work	Open source	Test Data	No. Params	Augmentation	Accur. %	Published in
VoxNet	Voxel	Theano/Lasagne	Y	MN 40	0.92 M	12 rot	83.00	2018
Principal Axle Descriptor	Voxel	Scikit-Learn	N	Own		N/A	93.18	2018
VoxelNet	Voxel	Tensor-flow	Y	KITTI		12 rot	89.35	2018
G3DNet	Point Cloud	Tensor-flow	Y	MN 40		N/A	91.00	2018
PointNet	Point Cloud	Tensor-flow	Y	MN 10	80 M	N/A	77.60	2017
FusionNet	Voxel + Pixel	Caffe	N	MN 40	118 M	60 rot	90.80	2016
NormalNet	Normal vector + voxel	Tensor-flow	N	MN 40	6.5 M	20 rot	88.80	2019
Variational A-encoder	Voxel	Theano/Lasagne	Y	MN 40	18 M	24 rot	88.98	2016
LightNet	Voxel		N	MN 40	0.3 M	12 rot	88.93	2018
3DShape Nets	Depth Map		N	MN 40	38 M	12 rot	77.30	2015
Object Net3D	Images		Y	Obj Net			98.90	2016

amount of points. It produces bounding boxes based on the features of the voxels. Finally, a combined approach, called FusionNet [47], uses both volumetric and pixel information to recognize 3D objects. A similar approach with face normals is used by NormalNet [47].

The rough overview in Table 6.5 shows disparate characteristics of pre-selected approaches, in particular the handling of augmentation. An accuracy in range of ninety percent based on ModelNet data sets can be expected at general. A ranking could be made based on more specific criteria. Apart of the extension of training data set to make it more representative, a trend for further improvement is not apparent, although few approaches were made open access in order to provide a platform for further development by developer community. Evidently, no method has yet been able to fulfil the promise of comprehensive semantically-rich building information model (BIM) creation [49]. The combination of two parallel, different approaches in one application to prevent overfitting may be a considerable, pragmatic option [39].

As practiced in many areas (autonomous driving, robotics, surveillance, or human–computer interaction), the real-time processing and recognition is an achievable option, but the expense for training can be hardly pre-estimated. The number of parameters indicates the necessary time for training. Important pre-requisite for a successful object recognition process is the availability of all relevant objects in the training database.

6.8 Application of Methods

In the variety of the object recognition methods presented here, it is necessary to select one or a few for practical use to generate the Digital Twin. Additional methods as illustrated in Fig. 6.10 must be also considered as support for not recognizable objects. Deep learning with 3D data has not been researched nearly as much as it has been with images. In comparison to object recognition methods based on images or depth maps (e.g. Object Net3D [40]), which achieve an accuracy of almost hundred percent for object detection, 3D point cloud data-based methods already have achieved higher accuracy in object localization, still require further development in area of object detection and provide higher potential for many applications (e.g. harsh environments).

By taking a closer look at the concrete results (see normalized confusion matrix in Fig. 6.11), it can be observed that the accuracy of 90% is probably more than sufficient in most cases, while considering the supporting measures (e.g. generation of the bounding box) that contribute to improving the overall results [46]. Based on ModelNet10 data set, for each object a high accuracy can be calculated. In contrast, for alternative (respectively wrong) objects maximal achieved accuracy is 0.12.

Convolutional Neural Networks (CNN) require the large size of datasets for training, which is the crucial challenge for industrial applications because creating a dataset for the training of specific industrial application is an expensive task. Main obstacles are both the recognition methods and the extremely time-consuming

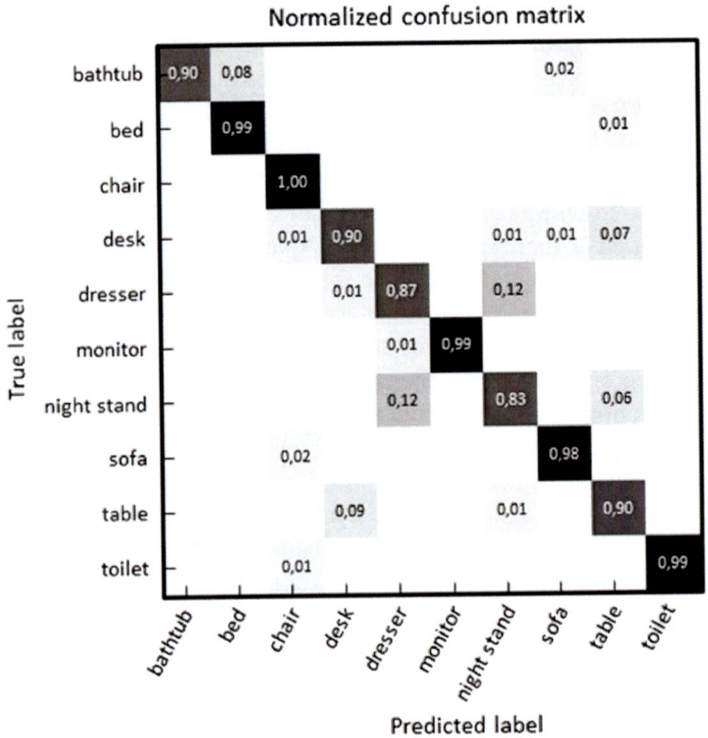

Fig. 6.11 Confusion matrix achieved on the ModelNet10 dataset [46]

handling of the huge volume of data. Therefore, it is necessary to prove several frameworks for 3D object recognition which promises the best results for investigated objects in the factory for different use cases. The future provision of a specific benchmark data set for "rough environments", which comprises typical industrial objects like machines, cranes, elevators, structural elements, pipelines, etc., by a standardization body or industrial consortium would also facilitate such a research [16].

Sometimes the dataset is not a constraint for applications, but the hardware for training. Typically, larger networks are expected to provide higher accuracy in CNN. Many researchers proposed methods to reduce input size, different layers and architectures to compress the size of the network or utilize low-computational power machines, some of them provide their method's source code for CPU utilizations [16].

Like most other object recognition architectures, voxel-based approaches are not inherently rotation-invariant. For this reason, a data augmentation technique needs to be implemented. During training, each model is rotated twelve to sixty times and trained on all copies. At test time, the output of the final fully-connected layer is pooled across several rotations of the input. In this way, voxel-based approach learns rotation invariance by sharing the same learned convolutional kernel weights across different rotations of the input voxel grid [31, 48].

Voxel grids still have a number of drawbacks. They lose resolution compared to point clouds, as several distinct points representing intricate structures will be binned into one voxel if they are close together. Voxel grids can yield unnecessarily high memory usage compared to point clouds in sparse environments, since they actively use memory to represent free and unknown space whereas point clouds contain only known points [31]. Finally, if an object is recognized, it must be put in the right position and proper orientation in the space.

Further important aspect to evaluate the representation type and the recognition method is the noise sensitiveness. Usually, noise reduction techniques are used to remove or reduce its effect in their methods and representations, avoiding to perform a specific analysis about noise sensitiveness. Some works ignore the noise concept in their study, just by using synthetic data and simulations. Others used noise as evaluation metric, where a method was considered robust and having good results when the final result had high classification rates in the presence of different noise levels [9].

The utilization of an adaptive grid can improve results in case of noise and occlusion. Adaptive grid sets its origin at the minimum x, y, and z values of the point cloud. Next, the grid size is adapted to the cloud dimensions. The leaf size is also computed in function of the grid size. Knowing both parameters, the grid is spawned, fitting the point cloud data. As a result, a non-cubic grid is generated. All voxels have the same size, but they are not necessarily cubic [50]. In such a way, non-desired filtering of local features can be prevented. It proves that considering real-world conditions is a matter of utmost importance when training these networks with synthetic datasets.

Further challenge is the inhomogeneity of point clouds which can occur for several reasons and have a critical impact on the results and reliability of the recognition process. A novel 3D object detection network called SARP-NET was developed in area of autonomous driving to solve the inhomogeneity of LiDAR point clouds and enhance the shape encoding capability. It encodes the sparse and inhomogeneous LiDAR point cloud in a more efficient and robust manner and appends an attention block to learn and deploy statistic 3D shape priors that are ignored in current one-stage 3D detectors. SARPNET outperforms the state-of-the-art one- stage and even two-stage methods for detecting objects in the Car and Cyclist classes while the performance of Pedestrian detection does not show a consistent enhancement. This happens probably because of that the pedestrians lack of discriminative features in 3D point cloud domain [51].

6.9 Discussion and Future Perspectives

As presented in the previous sections, object recognition is on the one hand subject of intensive research in field of computer vision with remarkable advancements. On the other hand, many practical applications have been deployed in the past years in wide areas. Subsequently, different methods (image-based and point cloud based) are maintained in parallel with their supporting development communities. Reliable video-based methods can be shortly expected. From this point of view, the current challenge is more to select and adapt an appropriate approach for object recognition which can be used and maintained in mid-term period rather than to develop a new one.

Based on the state-of-the-art in 3D object recognition area, some predictions of the direction yield that this area is taking in the next future [9]:

- Due to the even increasing number of methods employing feature-based representation, a trend of continuity with this representation type can be predicted, which would also include the addition of more features into the future proposed feature-based descriptors.
- Some works present feature descriptors combinations, which provide indicatives that the descriptors combination would be a possible solution to increase the objects' representation discriminative power. It looks like a solution to properly recognize less distinctive objects.
- For further penetration of more practical domains, a strong support by publicly available object databases is imperative. Several publicly databases were presented in the last few years for the 3D object recognition area. Thus, due to the increasing popularization of 3D data acquisition equipment, a more 3D object databases are expected to emerge in the next years. However, a standardization body which would collect objects from different vendors for sake of public utilization is still missing [27].
- Variables such as data, run time speed, computational resources, and required output all dictate the most suitable approach.
- Due to the advances in computation area, an increase in the research methods and techniques improving speed computation and reducing data dimensionality can be predicted. These improvements will emerge aiming the application of 3D recognition methods on real-time applications with several input information. That includes mobile devices, too.
- None of the methods provide a sufficient solution for specific practical issues: different working state (e.g. robots), occlusion, incomplete data sets, undesired reflections in scans on glass planes, etc. [27].
- An increase can also be inferred in the use of combined methods and learning models such as Deep Learning, due to the current increase in the use of those types of methods in several areas. Availability of supporting tools for learning operation can be expected.
- Majority of deep learning libraries are designed for desktop operating systems, but some libraries also provide support for mobile operating systems. However, some hardware manufacturer developed mobile devices for 3D support. The solutions of 3D-based methods can be deployed from almost all kind of devices, but low-computational or mobiles might take too much time in the recognition process [16].

– Lastly, there is a high probability of proposal of variations with well-known descriptors, using different similarity metrics, combinations and measures.

Finally, in spite of numerous object recognition methods which are being developed further almost concurrently, there is no ideal one with the best characteristics. Although promising results in addressing recognition tasks were achieved by Deep Learning methods, the underlying theory is not well understood, and there is neither clear understanding of which architectures should perform better than others. It is difficult to determine which structure, how many layers, or how many nodes in each layer are proper for a certain task, and it also need specific knowledge to choose sensible values such as the learning rate, the strength of the regularizer, etc. [52]. The design of the architecture has historically been determined on an ad-hoc basis.

The success of deep detectors relies heavily on gargantuan amounts of annotated training data. When the labeled data are scarce, the performance of deep detectors frequently deteriorates and fails to generalize well. In contrast, humans (even children) can learn a visual concept quickly from very few given examples and can often generalize well. Therefore, the ability to learn from only few examples, few shot detection, is very appealing. Even more constrained, zero shot object detection localizes and recognizes object classes that have never been seen before, essential for life-long learning machines that need to intelligently and incrementally discover new object categories [1].

This is also exemplified by the fact that no recognition system has consistently demonstrated graceful degradation as the scene complexity increases, as the number of object classes increases, and as the complexity of each object class increases [52]. In particular, CNNs struggle to generalize under challenging scenarios, like recognizing the variability and heterogeneity of the instances of elements belonging to the same category. That is exactly the situation in a machine hall where singular objects often distinguish from each other by some detail features.

Some of these difficulties are directly related to the input information, 2D-based methods still show a lack of robustness against strong lighting variations, for example. Merge techniques using both 2D and 3D information to overcome these problems can be an optimal way. A 2D labeling system for assigning categories to objects and describing them is used with a 3D feature descriptor in order to train a classification model. The experiments developed for the validation of the proposal show that the procedure developed has the capacity to recognize unseen objects based only on their 3D features, without the need to train a classifier using a highly similar instance of the object [17].

6.10 Conclusions and Outlook

The intention of this chapter is to present the object recognition methods which are usable in the built environment in order to close the gap between the real and virtual factory in the modern manufacturing industry as well as to make a pre-selection of the most suitable methods. The recognition and vision problem are highly transdisciplinary, spanning the fields of machine learning and decision making under uncertainty, robotics, signal processing, mathematics, statistics, psychology, neuroscience, human–computer interaction, databases, supercomputing, and visualization or graphics. The highly transdisciplinary nature of the problem is both an advantage and a disadvantage. It is an advantage due to the vast research opportunities it gives to the experienced vision practitioner. It is a disadvantage because the diversity of the field makes it all the more pertinent that the practitioner is careful and sufficiently experienced in identifying research that can advance the field [52].

Based on an overview of the recent object recognition publications, some of the major challenges facing the industrial users and emphasized some of the typical approaches attempted in the past few years for solving the recognition problem were presented.

The chapter begins with presentation of the needs and requirements of industry for the generation of a Digital Twin in a built environment. Based on this, the methodology for object recognition was drawn starting with challenges for object detection how to perform this task successfully. Basic classification of challenges is given. Methods for object recognition were classified according to their input in image, point cloud and video-based methods.

A pre-selection for point cloud-based methods is done and justified. Approaches for point cloud generation were classified according to the data source (monocular or binocular; image or video). While the recognition systems perform well in controlled environments, they are unable to generalize in less controlled environments. Therefore, an appropriate test base which represents the real conditions is necessary. Four of the most important test data sets were briefly introduced, although a data set for industrial equipment is still missed.

Process and method capabilities were also addressed. With regard to the quality of results, a particular attention was paid to accuracy of point cloud and entire recognition process. Generally, point clouds from laser scanners have high accuracy, and they are taken as ground truth to determine the accuracy of point clouds generated from monocular images, stereo images and video frames. Therefore, a laser is selected as the preferred data source.

Subsequently, methods for processing of point clouds were presented in detail. Need for preprocessing, mesh reconstruction, segmentation, sparse cloud modeling which is input for object recognition. Selected object recognition methods were compared based on few usability criteria. Some application constraints were assigned with object recognition methods. This chapter is concluded by discussion and explanation of future perspectives.

Selection of a suitable method for practical implementation is difficult because essential parameters differ greatly for individual approaches. Otherwise, oversupply and too little differentiation of results can be observed there. Considering further needs and derived goals a certain multi-channel implementation scenario can be drawn: simultaneous use of multiple devices (image, point cloud, or video) with collection of results and a subsequent decision making on deviations.

In term of research perspective, several directions can be predicted. Apart of the improvement of the accuracy, the trend to real-time execution will continue. The reduction of training period and expenses will play the crucial role. The "black-box" phenomenon of object recognition development and use must be addressed too: supporting the integration of code and documentation generation as well as explicitly linking knowledge base structure and meaningful content with the recognition application elements and code.

Finally, when considering the introduction and use of object recognition applications and templates, the adaptability and maintainability of developed applications and templates remains a point of concern in these turbulent times. Templates should cover the entire process as well as the most expensive steps (segmentation, object recognition) [53].

Looking from a higher level of abstraction can be said that effective and versatile automation capabilities of deep learning combined with large scale processing may be an adequate answer to the previously identified problems of today industry: data heterogeneity, massive scale, noisy data, and unstructured data. However, system interoperability still needs to be addressed in order to fully implement fully automated machine learning and deep learning workflows at large scale and deploy them seamlessly in the industry [54].

References

1. Liu L, Ouyang W, Wang X, Fieguth P, Chen J, Liu X, Pietikäinen M (2020) Deep learning for generic object detection: a survey. Int J Comput Vision 128(2):261–318. https://doi.org/10.1007/s11263-019-01247-4
2. Wognum N, Bil C, Elgh F, Peruzzini M, Stjepandić J, Verhagen WJC (2019) Transdisciplinary systems engineering: implications, challenges and research agenda. Int J Agile Syst Manage 12(1):58–89. https://doi.org/10.1504/IJASM.2019.098728
3. Ríos J, Mas Morate F, Oliva M, Hernández JC (2016) Framework to support the aircraft digital counterpart concept with an industrial design view. Int J Agile Syst Manage 9(3):212–231. https://doi.org/10.1504/IJASM.2016.079934
4. Quintana B, Prieto SA, Adán A, Vázquez AS (2016) Semantic scan planning for indoor structural elements of buildings. Adv Eng Inform 30:643–659. https://doi.org/10.1016/j.aei.2016.08.003
5. Zutshi A, Grilo A (2019) The emergence of digital platforms a conceptual platform architecture and impact on industrial engineering. Comput Ind Eng 136:546–555. https://doi.org/10.1016/j.cie.2019.07.027
6. Bondar S, Salem B, Stjepandić J (2018) Indoor object reconstruction based on acquisition by low-cost devices. Adv Transdisciplinary Eng 7(2018):113–122. https://doi.org/10.3233/978-1-61499-898-3-113
7. Geng Z, Bidanda B (2017) Review of reverse engineering systems—current state of the art. Virtual Phys Prototyping 12(2):161–172. https://doi.org/10.1080/17452759.2017.1302787
8. Poza-Lujan J-L, Posadas-Yagüe J-L, Simó-Ten J-E, Blanes F (2020) Distributed architecture to integrate sensor information: object recognition for smart cities. Sensors 20:112. https://doi.org/10.3390/s20010112
9. Carvalho LE, von Wangenheim A (2019) 3D object recognition and classification: a systematic literature review. Pattern Anal Appl 22:1243–1292. https://doi.org/10.1007/s10044-019-00804-4

10. Ma Z, Liu S (2018) A review of 3D reconstruction techniques in civil engineering and their applications. Adv Eng Inform 37(2018):163–174. https://doi.org/10.1016/j.aei.2018.05.005
11. Sommer M, Stjepandić J, Stobrawa S, von Soden M (2019) Automatic generation of digital twin based on scanning and object recognition. Adv Transdisciplinary Eng 10:645–654. https://doi.org/10.3233/ATDE190174
12. Sommer M, Stjepandić J, Stobrawa S, von Soden M (2020) Automated generation of a digital twin of a manufacturing system by using scan and convolutional neural networks. Adv Transdisciplinary Eng 12:363–372. https://doi.org/10.3233/ATDE200095
13. Bondar S, Ruppert C, Stjepandić J (2014) Ensuring data quality beyond change management in virtual enterprise. Int J Agile Syst Manage 7(3/4):304–323. https://doi.org/10.1504/IJASM.2014.065346
14. Ostrosi E, Stjepandić J, Fukuda S, Kurth M (2014) Modularity: new trends for product platform strategy support in concurrent engineering. Adv Transdisciplinary Eng 1:414–423. https://doi.org/10.3233/978-1-61499-440-4-414
15. Sommer M, Stjepandić J, Stobrawa S, von Soden M (2020) Improvement of factory planning by automated generation of a Digital Twin. Adv Transdisciplinary Eng 12:645–654. https://doi.org/10.3233/ATDE190174
16. Singh RD, Mittal A, Bhatia RK (2019) 3D convolutional neural network for object recognition: a review. Multimedia Tools Appl 78:15951–15995. https://doi.org/10.1007/s11042-018-6912-6
17. Rangel JC, Martínez-Gómez J, Romero-González C, García-Varea I, Cazorla M (2018) Semi-supervised 3D object recognition through CNN labelling. Appl Soft Comput J 65:603–613. https://doi.org/10.1016/j.asoc.2018.02.005
18. Xiao Y, Ma Y, Zhou M, Zhang J (2018) Deep multi-scale learning on point sets for 3D object recognition. Commun Comput Inform Sci 875:341–348. https://doi.org/10.1007/978-981-13-1702-6_34
19. Fathi H, Dai F, Lourakis M (2015) Automated as-built 3D reconstruction of civil infrastructure using computer vision: achievements, opportunities, and challenges. Adv Eng Inform 29(2):149–161. https://doi.org/10.1016/j.aei.2015.01.012
20. Liu Li, Ouyang W, Wang X, Fieguth P, Chen J, Liu X, Pietikäinen M (2020) Deep learning for generic object detection: a survey. Int J Comput Vision 128:261–318. https://doi.org/10.1007/s11263-019-01247-4
21. Zhao X, Ilieş HT (2017) Learned 3D shape descriptors for classifying 3D point cloud models. Comput Aided Des Appl 14(4):507–515. https://doi.org/10.1080/16864360.2016.1257192
22. Ghodrati H, Hamza B (2017) Nonrigid 3D shape retrieval using deep auto-encoders. Appl Intell 47:44–61. https://doi.org/10.1007/s10489-016-0880-1
23. Kamble K, Kulkarni H, Patil J, Sukhatankar S (2018) Object recognition through smartphone using deep learning techniques. In: 2nd international conference on soft computing systems, ICSCS 2018, vol 837. Communications in Computer and Information Science, pp 242–249. https://doi.org/10.1007/978-981-13-1936-5_27
24. Liu D, Cui Y, Chen Y, Zhang J, Fan B (2020) Video object detection for autonomous driving: Motion-aid feature calibration. Neurocomputing 409:1–11. https://doi.org/10.1016/j.neucom.2020.05.027
25. Ding X, Luo Y, Li Q, Cheng Y, Cai G, Munnoch R, Xue D, Qingying Yu, Zheng X, Wang B (2018) Prior knowledge-based deep learning method for indoor object recognition. Syst Sci Control Eng 6(1):249–257. https://doi.org/10.1080/21642583.2018.1482477
26. Wu Z, Song S, Khosla A, Yu F (2015) 3D ShapeNets: a deep representation for volumetric shapes. In: IEEE conference on computer vision and pattern recognition, pp 1912–1920. https://doi.org/10.1109/CVPR.2015.7298801
27. Salem B, Stjepandić J, Stobrawa S (2019) Assessment of methods for industrial indoor object recognition. Adv Transdisciplinary Eng 10:390–399. https://doi.org/10.3233/ATDE190145
28. Princeton Modelnet Repository. https://modelnet.cs.princeton.edu/#. Accessed 30 Sep 2020
29. Chang AX, Funkhouser T, Guibas L, Hanrahan P, Huang Q, Li Z, Savarese S, Savva M, Song S, Su H, Xiao J, Yi L, Yu F (2020) ShapeNet: an information-rich 3D model repository. https://arxiv.org/abs/1512.03012. Accessed 30 Sep 2020

30. Quadros A (2013) Representing 3D shape in sparse range images for urban object classification. Ph.D. thesis, The University of Sydney
31. Lehtola VV, Kaartinen H, Nüchter A, Kaijaluoto R, Kukko A, Litkey P, Honkavaara E, Rosnell T, Vaaja MT, Virtanen J-P, Kurkela M, El Issaoui A, Zhu L, Jaakkola A, Hyyppä J (2017) Comparison of the selected state-of-the-art 3D indoor scanning and point cloud generation methods. Remote Sensing 9:796. https://doi.org/10.3390/rs9080796
32. Rebolj D, Pučko Z, Čuš Babič N, Bizjak M, Mongus D (2017) Point cloud quality requirements for Scan-vs-BIM based automated construction progress monitoring. Autom Construct 84:323–334. https://doi.org/10.1016/j.autcon.2017.09.021
33. Dominguez M, Dhamdhere R, Petkar A, Jain S, Sah S, Ptucha R (2018) General-purpose deep point cloud feature extractor. In: 2018 IEEE winter conference on applications of computer vision, pp 1972–1981. https://doi.org/10.1109/WACV.2018.00218
34. Nguyen A, Le B (2013) 3D point cloud segmentation: a survey. In: 2013 IEEE 6th international conference on robotics, automation and mechatronics (RAM). https://doi.org/10.1109/RAM.2013.6758588
35. Zhang Le, Sun J, Zheng Q (2018) 3D point cloud recognition based on a multi-view convolutional neural network. Sensors 18:3681. https://doi.org/10.3390/s18113681
36. Liu X, Lu Y, Wu T, Yuan T (2018) An improved local descriptor based object recognition in cluttered 3D point clouds. Int J Comput Commun Control 13(2):221–234. https://doi.org/10.15837/ijccc.2018.2.3010
37. Zhang W, Su S, Wang B, Hong Q, Sun L (2020) Local k-NNs pattern in omni-direction graph convolution neural network for 3D point clouds. Neurocomputing 413:487–498. https://doi.org/10.1016/j.neucom.2020.06.095
38. Wang K, Kim M-K (2019) Applications of 3D point cloud data in the construction industry: a fifteen-year review from 2004 to 2018. Adv Eng Inform 39:306–319. https://doi.org/10.1016/j.aei.2019.02.007
39. Sommer M, Stjepandić J, Stobrawa S, von Soden M (2021) Automated generation of a digital twin in manufacturing for a built environment using scan and object detection. J Indus Inform Integr XX (in press)
40. Pătrăucean V, Armeni I, Nahangi M, Yeung J, Brilakis I, Haas C (2015) State of research in automatic as-built modelling. Adv Eng Inform 29(2):162–171. https://doi.org/10.1016/j.aei.2015.01.001
41. Chen J, Fang Y, Cho Y (2018) Performance evaluation of 3D descriptors for object recognition in construction applications. Autom Construct 86:44–52. https://doi.org/10.1016/j.autcon.2017.10.033
42. Brock A, Lim T, Ritchie JM, Weston N (2016) Generative and discriminative voxel modeling with convolutional neural networks. 3D Deep Learning Workshop at NIPS 2016. http://3ddl.cs.princeton.edu/2016/papers/Brock_et_al.pdf. Accessed 12 June 2019
43. Qi CR, Su H, Mo K, Guibas LJ (2017) PointNet: deep learning on point sets for 3D classification and segmentation. In: Proceedings of the IEEE conference on computer vision and pattern recognition, Honolulu, HI, USA, 21–26 July 2017
44. Maturana D, Scherer S (2015) VoxNet: a 3D convolutional neural network for real-time object recognition. In: IEEE international conference on intelligent robots and systems, 2015-December, 7353481, pp 922–928. https://doi.org/10.1109/IROS.2015.7353481
45. Zhou Y, Tuzel O (2018) VoxelNet: end-to-end learning for point cloud based 3D object detection. In: The IEEE conference on computer vision and pattern recognition (CVPR), pp 4490–4499. https://doi.org/10.1109/CVPR.2018.00472
46. Zhi S, Liu Y, Li X, Guo Y (2017) LightNet: a lightweight 3D convolutional neural network for real-time 3D object recognition. In: Eurographics workshop on 3D object retrieval, pp 9–16. https://doi.org/10.2312/3dor.20171046
47. Hegde V, Zadeh R (2019) FusionNet: 3D object classification using multiple data representations. In: 3D deep learning workshop at NIPS 2016. https://stanford.edu/~rezab/papers/fusionnet.pdf

48. Wang C, Cheng M, Sohel F, Bennamoun M, Li J (2019) NormalNet: a voxel-based CNN for 3D object classification and retrieval. Neurocomputing 323:139–147. https://doi.org/10.1016/j.neucom.2018.09.075
49. Czerniawski T (2020) Fernanda Leite, (2020) Automated digital modeling of existing buildings: a review of visual object recognition methods. Autom Construct 113:103131. https://doi.org/10.1016/j.autcon.2020.103131
50. Garcia-Garcia A, Garcia-Rodriguez J, Orts-Escolano S, Oprea S, Gomez-Donoso F, Cazorla M (2017) A study of the effect of noise and occlusion on the accuracy of convolutional neural networks applied to 3D object recognition. Comput Vision Image Understand 164:124–134. https://doi.org/10.1016/j.cviu.2017.06.006
51. Ye Y, Chen H, Zhang C, Hao X, Zhang Z (2020) SARPNET: shape attention regional proposal network for liDAR-based 3D object detection. Neurocomputing 379:53–63. https://doi.org/10.1016/j.neucom.2019.09.086
52. Guo Y, Liu Y, Oerlemans A, Lao S, Wu S, Lew MS (2016) Deep learning for visual understanding: a review. Neurocomputing 187:27–48. https://doi.org/10.1016/j.neucom.2015.09.116
53. Kuhn O, Liese H, Stjepandić J (2011) Methodology for knowledge-based engineering template update. In: Cavallucci D, Guio R, Cascini G (eds) Building innovation pipelines through computer-aided innovation. Springer-Verlag, Berlin Heidelberg, pp 178–191. https://doi.org/10.1007/978-3-642-22182-8_14
54. Dekhtiar J, Durupt A, Bricogne M, Eynard B, Rowson H, Kiritsis D (2018) Deep learning for big data applications in CAD and PLM—Research review, opportunities and case study. Comput Ind 100:227–243. https://doi.org/10.1016/j.compind.2018.04.005

Chapter 7
Data Quality Management for Interoperability

Josip Stjepandić and Wjatscheslaw Korol

7.1 Introduction

Data becomes an even more important constituent of our daily lives and an essential asset, driver and fuel for mature and new industries. While data is becoming a crucial importance as a strategical resource, the quality of the data that is used during business processes will impact the business outcome today and tomorrow [1]. As a fundament of the digital economy, data must have a high quality. For the data quality, the same semantic understanding is transferred from the production of physical goods to the management of data [2]. The quality of data, defined by the objectively measurable degree to which the data properties fulfill requirements, can have a tremendous impact on the businesses [3]. Subsequently, data can be seen as a raw material for the production of information products through an information production process [4]. This also fits the definition of information as the "processed data". Since information products reproduce itself almost recursively, the terms data and information can only be distinguished in the basic process chains and used synonymously further [5]. Data quality is usually assumed [6]. It becomes a topic once the data quality falls below a certain threshold [7].

Research and industry reports continuously discover that huge efforts are spent to improve the quality of the data being used in many applications, sometimes even only to understand the quality of auxiliary data, in order to preserve a proper function of information systems. Data quality is defined as a context-dependent, multidimensional property and expresses the fitness for use of certain data for a user in a specific context. The inherent context dependence of the data quality emphasizes that the

J. Stjepandić (✉) · W. Korol
PROSTEP AG, Dolivostr. 11, 64293 Darmstadt, Germany
e-mail: josip.stjepandic@prostep.com

W. Korol
e-mail: wjatscheslaw.korol@prostep.com

J. Stjepandić et al. (eds.), *DigiTwin: An Approach for Production Process Optimization in a Built Environment*, Springer Series in Advanced Manufacturing,
https://doi.org/10.1007/978-3-030-77539-1_7

requirements for data elements depend on the intended use and only the user can decide for himself whether the data objects are usable for him or not. The context can include, for example, the data-using business process, country affiliation, applicable regulations, time of data usage, data-processing application or the business process role of the data user [8].

Considering the variety of business views, use cases, properties, or simply the specificities of the systems being evaluated, the quantitative assessment of the data quality can become an extremely difficult task which can hardly provide clear results [9]. In general, the importance of data quality increases the more complex the business processes become and the more applications and interfaces have to interact [10]. Data translation mostly impairs the data quality [11] and, therefore, cause for several issues.

In addition, there are many data quality dimensions which, in combination, express the data quality (Fig. 7.1): timeliness, credibility, reliability, interpretability, operability and sufficiency accompanied with data quality attributes. The quality of a data object can be measured by checking certain properties of the values of the data elements it contains [13]. If the data elements of a data object have all the required properties, its quality is perfect. Properties of data (e.g. certain value is mandatory within a range, completeness of a data element) can be formulated as business rules and thus checked. There are structural (how must a data object be structured?) and operational business rules (how do values for individual data elements have to be set?) [8].

Data Quality Management is a subordinated area of data management, which, as part of company-wide information management, aims to make optimal use of data in the company. Data quality is never an end in itself. As a typical supporting topic, it gets its meaning through a process chain, mostly as a pre-requisite for a highly performant interoperability [14, 15]. Subsequently, the term digital thread refers to a communication framework [16]. Companies can only offer digital services, open up new business opportunities or make processes between companies more efficient, if data on customers and products, but also context information on whereabouts, preferences, and billing are available in high quality [1].

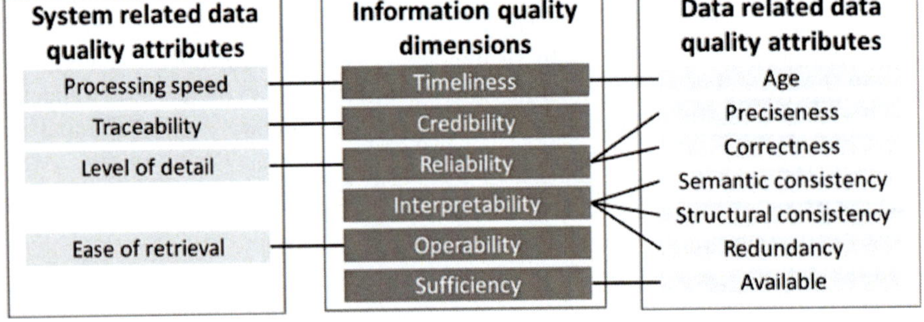

Fig. 7.1 Data and information quality model [12]

Like other business assets, data also has its own lifecycle that needs to be managed in a suitable way to continuously ensure its purpose and need. The lifecycle of a data object begins with its planning and specification, through the creation during the execution of business processes, and ends with their archiving or deletion [8]. It is important to say that the life cycle of the data is always longer than the life cycle of the product concerned due to the regulatory rules. This fact has a crucial impact to Digital Twin because all components must be adjusted with one another in order to enable seamless interaction. A high data quality is a pre-requisite for this.

The given definition of the data lifecycle corresponds to the understanding of the product lifecycle, which also begins with the first product requirement and not with the current product representation. Established reference models of Product Lifecycle Management (PLM) describe several process stages for planning and implementing the entire product life cycle [17]: planning the product portfolio, designing the product, planning the production process, supplying the end customer with the product, providing service and support services, and, finally, product disposal and recycling. In era of model-based definition and processes, the data management including data quality management is a function within a powerful commercial PLM system [17, 18]. Thus, one of the aims of PLM is the provision of data in a sufficient quality for downstream processes.

The structure of this chapter reflects this aim. In Sect. 7.2, Digital Thread and its supporting concept are briefly introduced. Data quality classification with respect to its dimensions and related standards is discussed in Sect. 7.3. Subsequently, achievements related to the issue of data quality metrics in manufacturing industry is introduced in Sect. 7.4. Section 7.5 showcases the achievements of data quality in industrial applications of design and manufacturing for various industries. A discussion chapter in Sect. 7.6 gives insight into benefits and gaps of current applications of data quality as well as future directives. Finally, an outlook is given in Sect. 7.7 with respect to the future importance of data quality from business process perspective.

7.2 Digital Thread

The fundamental vision of integration in manufacturing industry supposes a seamless flow of information in all product lifecycle stages from the first product idea until disposal and recycling. This vision is expressed by three layers as illustrated in Fig. 7.2 which can be seen as structural constituents of Digital Twin.

Data integration is a widely requested 'digital data' lever for digital transformation. It describes a product (born digital) holistically with (1) domain-specific application models for example, mechanical, software, simulation or cost models. It demands cohesive communication in the (2) supply chain based on business data streams with partners, in joint ventures and across factory plants [19]. It finally realizes (3) a fusion between up- and downstream in the entire lifecycle, where digital aspects of the product solely are used as engineering, manufacturing and service

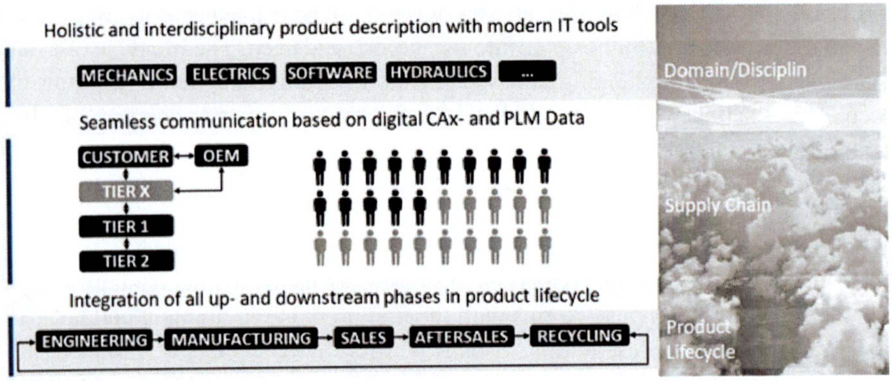

Fig. 7.2 The integration vision in the automotive industry [20]

bridges [20]. Although all three layers have its particular importance for Digital Twin, the product lifecycle layer (3) prevails due to the volume and complexity.

The connection between the real asset and the development and planning models that describe its history is known as a digital thread [16]. Like a data highway, it connects the information of a real product instance across processes and IT systems. On the one hand, this enables all data from the lifecycle of the product instance or the real asset to be brought together and thus forms the basis for the creation of Digital Twins. Without digital thread, the Digital Twins could be recreated manually, but it would be difficult or impossible to keep them up to date. On the other hand, the traceability along the digital thread makes it possible to track and monitor decisions in development and production and to identify potential for optimization with the help of the operating data [21].

Product data is an umbrella term that includes many different types of information: PDM, CAx, planning, inspection data [17]. Usually only certain data is necessary at sequences (gates) for assessment in downstream processes [18]. Large and comprehensive functionalities, such as provided with modern CAx systems are not mandatory. In fact, the number of consumers of CAD data in extended enterprise is about a factor of ten at least higher than data creators in engineering. The use of powerful CAx systems however in downstream processes such as purchase, production, assembly and quality assurance needs to be scrutinized in the course of an efficient product creation [22]. Certain level of data quality and level of detail (filter) is always supposed.

7.3 Data Quality Classification

To properly place data quality, clear classification criteria are necessary. Classification can be based on several criteria (environment, organization, purpose, status). For our purpose, its classification relative to the afore-mentioned dimensions (Fig. 7.1) needs to be understood.

7.3.1 Data Quality Dimensions

Ensuring the timeliness of data processing requires the ability to acquire, transfer, process, transform and use the data in required time. It comprises the temporal capability of the virtual representation of entities from the real world, based on the data quality attributes age and processing speed. Age refers to the time passed since a change of the measured value in context of change. Processing speed is a system-related attribute measuring the interval from data collection to information provision [12]. Timeliness expresses the time expectation for accessibility and availability of data. As the value of data can rapidly decrease over time, the computing architecture needs to perform all the calculation and communication almost on the fly with the data that was recently provided.

Data credibility expresses the level of certitude to which the good faith of a source of data can be relied upon to facilitate its reuse in order that the data really represents is what the data is supposed to represent, and vice versa [23]. In other words, data credibility indicates the confidence of the Digital Twin users supported by the trustworthiness of the provided data. It is based upon the consistency with other evidence. It fosters the willingness to rely on the data with the goal to increase the intensity of data usage [12].

Data traceability is the ability to track a data construct back to the construct it was derived from as a more concrete instantiation. This can be ensured by metadata that track information provenance, for instance implemented in form of a data pedigree (effects of attributional qualities of a source). A pedigree is a list of ancestors with some attribution of purity of the lineage [24].

Data reliability is the degree to which prior historical reports from a source have been consistent with fact. Reliability includes a notion of dependability, that the data will be produced, and attain some level of accuracy and precision [24]. The data quality attribute correctness binarily differentiates data considered 'correct' or 'incorrect'. This requires that all data values for a business attribute must be correct and representative of the attributes. Preciseness is a data-related attribute measuring inaccuracy on data item level, while level of detail is determined by system limitations. While this is trivial for numeric values, other data types need to be translated into a numeric representation that allows for deviation measures, or special distance metrics need to be applied [12].

The information provided from Digital Twins must be interpretable for the users. Interpretability can be assessed at two different levels: by examining models (heuristic approach) or representations (mainly with user-based surveys). In the former case, simple measures can be used to compare several models of the same type, such as the number of rules and terms in decision rules or the number of nodes in decision trees. If models differ, then this comparison is not that obvious and other heuristics have been proposed [25].

The provided data should be free of unintentional duplicates, as expressed by the data quality attribute redundancy. In contrast, an intentional redundancy can be useful and improve process reliability [4]. Besides, defined rules need to be satisfied, as expressed by semantic consistency and structural consistency. Semantic consistency describes consistency of the meaning of data, achieved through unified definitions, labels assigned to real-world objects, and vocabulary [12]. Semantic consistency is important for the mathematical data quality of CAD data. Structural consistency refers to technical specifications of structure and format and impacts the organizational quality of CAD data [4].

Data operability describes a level of data record ability to be used directly, without additional processing (translation, filtering): how the information consumer interacts with the Digital Twin. The quality attribute ease of retrieval is assessed on a scale from 'inaccessible to user' to 'machine readable and ready as input for analysis software' [12]. It belongs to the specific data operability challenges such as semantic duplicates, data fusion or information extraction.

Data sufficiency is related to the amount of information provided to fulfil a certain purpose, e.g. a complex assembly feature. Insufficient data yield unstable models. The attribute availability of a dataset describes whether mandatory data items are available, e.g. in case of data replication [12].

From the engineering perspective, the data quality dimensions can be roughly classified in two groups: technical and organizational. Technical dimensions are primarily related to the capabilities of product and organizational to the capabilities of process. In order to check the data quality by appropriate tools, technical dimensions are checked for singular entities. In contrast, organizational dimensions are checked for data structures.

7.3.2 Related Standards

The primary purpose of standards is to define compliance clauses in a way that vendors can claim compliance to differentiate their offering from those that are not compliant. Compliance with an international, industry or proprietary standard is often a legal pre-requisite for a supplier to get a contract with his customer [19]. While a specific Digital Twin standard, which would define the necessary data quality requirements, neither exists yet nor can be expected in the next future, the existing standards related to data quality will be shown here.

ISO 8000 is a set of data quality management standards developed by ISO TC 184/SC4/W G13 [26]. The committee's mission is the development of standards for the exchange of complex data in an application neutral form to provide data portability and long-term data preservation in an environment where the life cycle of software applications used to capture and manage data is but a fraction of the life cycle, if the data itself lie in focus of this committee. ISO 8000 defines which characteristics of data are relevant to data quality, specifies requirements applicable to those characteristics, and provides guidelines for improving data quality. It deals with Master Data, Transactional Data, Referenced Data and Engineering Data. ISO 8000 standards can be applied to manufacturing processes defined in IEC 62,264 (Enterprise-control system integration). Manufacturing processes defined in IEC 62,264 are restructured according to processes of ISO 8000-61. Each process consists of purpose, outcomes and activities. Achievement of each process can be confirmed with work products [27].

For sake of Digital Twin, the data quality management process reference model of ISO 8000-61 [28] and process assessment of ISO 8000-62 [29] are of a particular importance. ISO 8000-61 specifies the processes required for data quality management. Each process is defined by the purpose, outcomes and activities that are to be applied for planning, controlling, assuring and improving data quality. It comprises also the data-related support and resource provision. The processes are used as a reference model in assessing and improving data quality management. The implementation cycle is based on the 'Plan, Do, Check, Act'-cycle defined in ISO 9001. Based on 8000-61, ISO 8000-62 identifies those elements of the maturity model that exist in other standards and specifies additional elements of the maturity model. ISO 8000-62 provides guidance on assessing the maturity level of an organization and derives organizational process maturity level rating from process profiles. The assessing of the organizational maturity level for data quality management conveys how well the organization is fulfilling the requirements identified by the process reference model specified by ISO 8000-61 [27]. ISO 8000-62 specifies six maturity levels and process profiles to indicate when organizations have achieved each of the maturity levels. ECCMA (Electronic Commerce Code Management Association) has developed a series of compliance certificates for individuals, organizations and their software applications and data services [29].

IEC 62,264-1:2013 defines the functions of an enterprise involved with manufacturing and the information flows between the functions that cross the enterprise-control interface in order to improve integration regardless of the degree of automation. Global acting companies are very interested in it because it unifies and merges different IT methods and enables robust, easy-care integration solutions to be achieved in the long term. This standard is important for both manufacturers and users and system integrators. It offers a uniform terminology for corporate IT and control systems as well as a number of concepts and models for the integration of corporate functions. The technical solution is determined by the uniform modeling of the interfaces between the corporate functions and control functions. The main concepts are object modeling and the modeling languages UML and XML [30].

Among the industry standards for data quality, the guidelines that have been developed by the German and international automotive industry since 1980 have achieved a widespread use. These activities were driven by increasing complexity of product models which are described by thousands of CAx and PDM tools in the automotive supply chain. The difficult data exchange was caused by not powerful interfaces and low data quality [31].

The main purpose of the widespread VDA (German automotive association) recommendation 4955/2 is to improve the collaboration of project partners during the product development phase [32]. This guidance helps to reduce remastering times and costs of CAD processes through the exchange of information and experience, the definition of cross-company, common data quality criteria, the provision of both CAD system-neutral test programs and repair aids as well for the frequently used systems or system pairings. The recommendation is divided into the following areas:

- geometrical data quality,
- organizational data quality,
- recommendation for agreement among data exchange parties, and
- recommendation for a proper extent of CAD models.

This recommendation is written in such a level of detail like an implementation guide for the development of check tools. Although the fulfillment of quality criteria is checked using proprietary system functions, the criteria are formulated neutrally in order to preserve comparability and system independence of the results. The translation and check software usually are certified by VDA against this recommendation, whereby the various check tools are compared by using different test models from practice in order to guarantee the reliability and consistency of the test results. An example for consistency issue in a CAD model with a virtual gap in solid that really does not exist is shown in Fig. 7.3.

Similar standards were established in further countries with automotive industry. Then, these have founded the association SASIG (Strategic Automotive product data Standards Industry Group) which has taken the further development of data quality standards. The result was "Product Data Quality Guidelines for the Global Automotive Industry" [33], which has adopted and extended the previous works of VDA. The last version of this document entails CAD, CAE, PDM and inspection data. It contains some suggestions for project management, communication and know-how for better CAD model quality. However, these suggestions are generic and cannot be applied directly. A direct implementation in a check tool is not known yet [33].

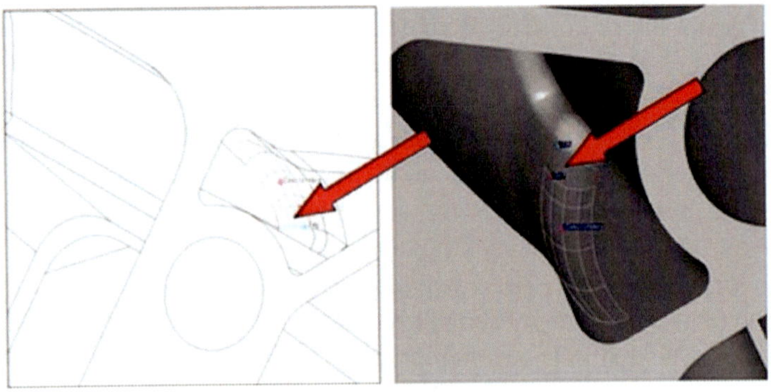

Fig. 7.3 Exemplary data quality problem: consistency [31]

7.4 Data Quality Metrics

A metric, or Key Performance Indicator (KPI), is a quantifiable attribute of an entity or activity that helps to describe its performance [18]. This can be measured to help, manage and improve the entity or activity. In term of data quality, it represents a set of attributes which describe the data quality in a sufficient way. From a quality perspective, two moments have importance during the data's lifetime: the moment it is created and the moment it is used. The quality of data is fixed at the moment of creation or change. However, data quality is usually not assessed until the moment of use. If the quality looks low, users typically attempt by working around the data or correcting errors themselves. Therefore, quality of product models needs to be continuously controlled in engineering workflow, especially in systems based on downstream data.

Model quality impacts not only the model accuracy and modifiability but also the changeability of the whole engineering systems. Careful and thorough model verification facilitates effecting product model quality. Verifying product models and designs manually is a tedious and time-consuming process [19]. By automating parts of the verification process, e.g. by using intelligent templates for check tools [34], benefits can be achieved in the time frame and end results of the verification.

The metrics presented below belongs primarily to CAD data as the main input for Digital Twin. A 'one size fits all' set of metrics is not a solution and that assessing data quality is an on-going effort that requires awareness of the fundamental principles underlying the development of subjective and objective data quality metrics. The main dimensions for the metrics are shown in Fig. 7.4. The dimensions contain almost universally agreed model quality dimensions described in Sect. 7.3.1. These dimensions should be the basic principles when assessing product model quality. In addition, often used quality dimensions are accessibility or reachability of data. Considering the downstream processes, modifiability and reusability are the paramount dimensions. However, for all-around measurement for quality of configurable product models, even more additional dimensions are needed [35].

Simplicity of product models defines the topological structure built of a few simple and understandable elements. Simple models facilitate consistency in the modeling

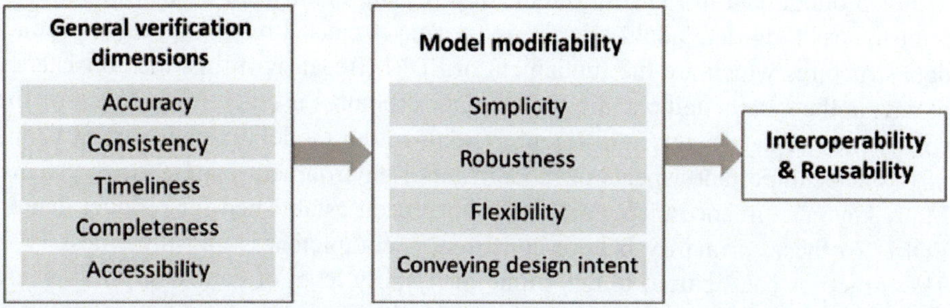

Fig. 7.4 Model verification metrics [35]

system in order to reduce causes of possible instability. Model instability would cause huge additional efforts to repair the model. Robustness of a model describes the resistance to error while being modified. It is a prime indicator for overall model quality as it is a result of minimizing error and quality issues from models [35]. Basically, the approach to create robust models means to create simple models with as simple features as possible. Referencing increases the robustness if detail feature shows to the basic feature [31].

Flexibility expresses the range of reachable states depending on the time and cost required to change state. Flexible systems are built for a set of reachable states which are predefined during the engineering process [35]. Interoperability describes how a master model can be accurately transferred from one format to another, e.g. CAD to CAD or point cloud to CAD. Interoperability can be efficiently supported to create and follow common methodology for modeling in all systems. Interoperability is a prerequisite for seamless downstream processes in Digital Twin [31].

Reusability describes good model reuse derived from the structures and references of the model. Reuse can be performed at different levels from utilizing library component—as practiced in our solution—to using existing designs of similar properties. As the modification to the models in the case of reuse is frequent, the quality of reusable models must be higher than normal as they need to reliably allow for modifications while maintaining the original design intent. Design intent and rationale are always needed for a model to be reusable [35]. Therefore, Digital Twin requires a native model in similar quality as a manually generated CAD model following a method. Conveying design intent is important for some newer engineering processes such as Model-based Design (MBD), Model-based Engineering (MBE) and Knowledge-based Engineering (KBE). Simplicity helps to convey the intent of the original design.

On a morphological level the model quality can be quite effectively quantified even with CAD environment native tools. Different type of geometry checks and identification of topological errors is usually even built-in the software to make sure the software works as it should [35]. Such functions are often collected in a specific data quality check module. Such modules can be easily controlled by using pre-defined profiles for specific purposes [34]. The most important metric on data quality is the distribution of error rate during a period, e.g. project duration [4].

For product data in PDM different rules of data quality are used which mostly refer to structural data quality dimensions. This is caused by configurable product data structures which are the fundament of PDM. Because of this, the consistency of data in the system and planning of the data structures is crucial. Furthermore, the consistency of data between different systems in the modeling environment is one of the most important aspects in modern multi-environment engineering systems. Many key performance indicators (KPI) have been established for product data in PDM. To these, primarily belong consistency, completeness and timeliness [17]. These metrics can be used to determine the quality level of data sets, but they are not as suited for evaluating and verifying the quality level of single product data instances. Furthermore, some metrics or parameter thresholds need to be used to correctly evaluate if product data is incomplete or not [35].

7.5 Practical Examples

The framework of the practical application is usually formed by the standard steps according to ISO 9000: quality planning, control, testing, and improvement. The focus is set here to control and testing. Practical management of data quality consist of methods to ensure required data quality and their implementation in process chains—primarily by deploying suitable data quality check tools. Good product data quality means providing the right data to the right task at the right time [33]. Data provision is realized by workflows in modern PDM systems. Data quality methods ('Design for Quality') help users in usage of IT systems (e.g. CAD) to ensure the quality of their work from the perspective of various stakeholders. It consists of approaches for manual techniques for assessment of CAD models, identification and repair of typical model errors, as well as application of modules and tools for interactive improvement of models [31].

The quality assurance should occur in each phase when data is changed. It gets its particular importance during the detail design, when both the amount of new generated data and the frequency of changes are high. Therefore, the practical data quality assurance is focused either on the data creation period or its phase of exploitation. In this section, three practical examples are presented which belong to Digital Twin: the use of a commercial tool in the design phase with focus on the methods and training (Sect. 7.5.1) [36], the conception and implementation of a knowledge-based check tool for downstream application in manufacturing (Sect. 7.5.2) [37], and the control of CAD data migration process by using a CAD quality tool (Sect. 7.5.3) [4].

7.5.1 Design

The study presents the use of a standard, commercial module of SolidWorks (SolidWorks Design Checker—SWDC) in a representative case study of modern Model Quality Testing tools (MQT) [36]. SolidWorks is a leading CAD system, widely used in manufacturing industry which is also applied as authoring system for factory models in our DigiTwin project [37]. SWDC can identify and sometimes repair data errors that could affect the simplification, interoperability, and reusability of CAD models.

This study has investigated the usefulness of this check tool as an assessment mechanism both for instructors and for self-evaluation. SWDC integrates the following modules: build checks, check active document, check again existing file, and learn check wizard. By mapping the Build Check requirement of SWDC against the CAD quality criteria available in the literature, the main conclusions can be drawn that SWDC only covers lowest semantic level quality criteria, and is designed for intensive use to maintain consistency across documents [36].

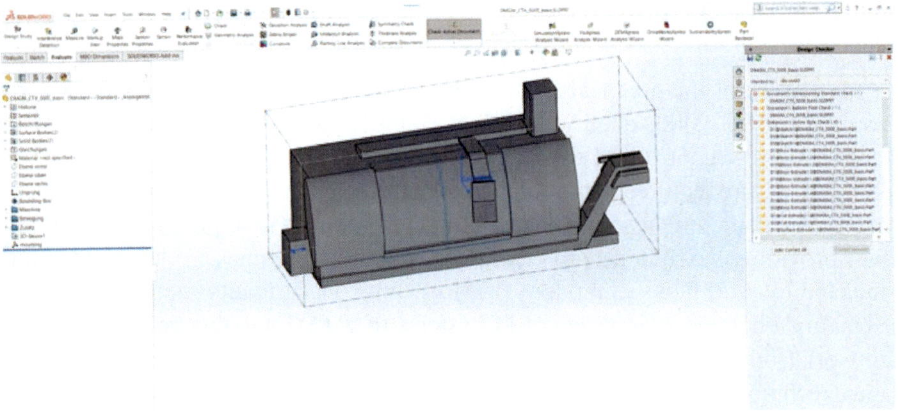

Fig. 7.5 Usage of SolidWorks Design Checker—SWDC in context of digital twin

This study provides an exhaustive insight how to map requirements to quality dimensions. Issues arise on detection of constraints that are repetitive but not incompatible. In general, the structure of the application is designed for intensive use and to maintain consistency across vast amounts of documents. Most criteria implemented in SWDC are aimed at verifying settings, and thus intended to ensure the semantic correctness of the CAD model [36].

Two additional observations are included: SWDC will repair certain errors, but others (not all) will only be identified, and SWDC partially overlaps with the built-in checking capabilities of the CAD application, which can sometimes perform better than the MQT [36]. The findings include the insight that the product data quality can be achieved by design rather than by MQT tools. This is practised in our development of Digital Twin (Fig. 7.5) [37].

Finally, although MQT is considered solved by a number of scholars, it remains an open practical problem, as new quantitative metrics require the design of new application programming interfaces that transform current MQT tools into mechanisms to assess higher semantic level quality aspects [36].

7.5.2 *Manufacturing*

Different model users in various stages of the manufacturing process have varied requirements in terms of model quality. To handle this diversity of models, a study proposes a knowledge-based MBD part model quality analysis system and its implementation technologies to analyze and test the quality of models from the perspective of different model-used stages. Alternatively, such a system would need to create partial models by using entity filters. It is fully integrated into CAD system Siemens NX and, therefore, uses system functions for its operations. The system decomposes

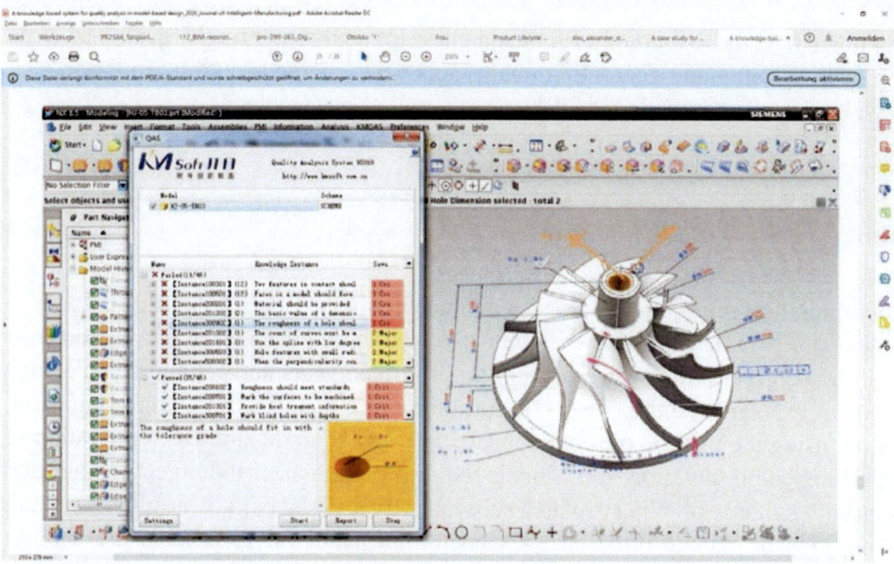

Fig. 7.6 Data quality analysis in model-based design [38]

the MBD part model into model definition instances (MDI) as well as the relationship between them for model quality verification with the help of relevant model quality knowledge. The model quality knowledge from different model-used stages enables the system to analyze quality defects from the respective perspective of model-used stages in MBE [37]. After the analysis is finished, model quality issues and corresponding modification suggestions are written into report.

Based on a knowledge-management system with its own data quality data model, this tool acquires new model quality knowledge into the knowledge base. The rule-based framework of knowledge representation facilitates the functional extension of this system. The applicability and expressive power of the rule-based framework is enhanced by object chain rules and parameter table rules (Fig. 7.6). In practice, with the help of the system, the knowledge derived from various model-used stages in MBE is collected, stored, and reused in downstream processes [38]. The unknown model defects are represented by new knowledge objects.

Apart of the detection of the model quality defects, this system provides modification suggestions in the design stage. However, the quality defects mostly need to be addressed manually, which results in a lot of repetitive work. Hence, new quality defects may occur for the wrong modification operations. Moreover, only the qualitative analysis is provided in this system [38]. The most important drawback is the missing comparing functionality. It is difficult for the analysis results of different models to be compared directly. Besides, for different quality analysis requirements from various models, the analysis scheme has to be created manually [38]. It is tedious in case of an assembly model which contains of a large variety of part models.

For a frequent use in generation of a Digital Twin three improvements should be made: the procedure or workflow to automatic quality defects modifications based on knowledge, the technology for the quantitative analysis of model quality, and the approach to automatic generating quality analysis scheme according to the model, as well as relevant knowledge [39].

7.5.3 *Data Migration*

Data migration refers to a large system change which includes the translation of a huge quantity of data which were often collected during decades within a short time period. Such migrations happen every ten to fifteen years and set a huge challenge for the organization and the users. Remastering of a factory with all the equipment, e.g. by using the method presented in Chap. 6, can be understood as a migration of the factory representation, too. This is valid in particular, if the object recognition does not recognize all objects which subsequently must be built by a singular feature recognition or manual remastering. Generation of Digital Twin must preserve an option to remove a system by another.

In such a case, a huge change arises in the customer process of almost each internal and external supplier, because they are forced to keep the current process running and to ramp-up the new process. This procedure includes many CAD translation steps which are principally not beneficial for good data quality. The challenge is ensuring an appropriate level of data quality to make sure that all translating processes are successful [4].

The translators in modern CAD systems show a similar level of performance and robustness as known from previous benchmarks and long-term experience [4] and, therefore, all models can be transferred lossless without exception to SolidWorks. However, it appears that in some cases automatic healing algorithms have slightly adjusted the geometry to satisfy the continuous condition [39]. It could not be predicted definitely to what extent of additional problems in further processing this will lead [4].

Further comparisons in the model properties like center of gravity, moments of inertia, as well as cloud of points, were also executed systematically (Fig. 7.7). All values remained within the allowed tolerances and showed no abnormalities what indicated a mature and stable and reliable process [4].

7.6 Discussion and Future Perspectives

Despite of its huge achievements described in the use cases above, data quality is still subject to intensive research and practical improvements. The valuation of data quality can be facilitated in two directions: operational control of data quality as autonomous procedure (supported by an application) and data quality as inherent component of the business process as a whole.

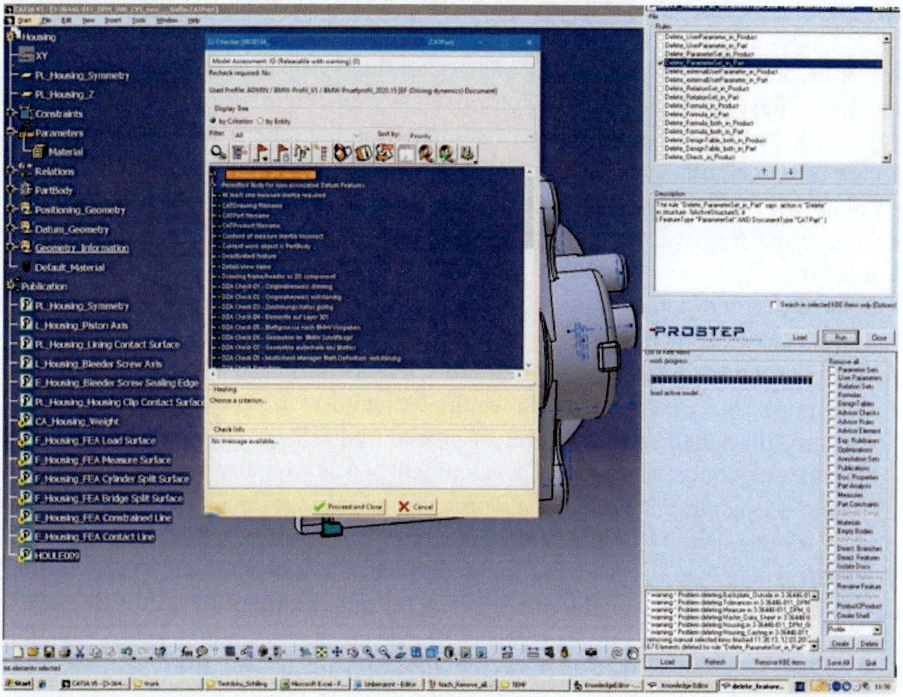

Fig. 7.7 Results of data quality check of migrated data

When considering the operational control of data quality, a number of developments have put data quality on a stronger footing over the last decade. In particular, the development and uptake of data quality dedicated design methods, the uptake of data quality assessment functionality in most major CAD and PDM systems, and the development of supporting methodologies as well as check tools have strongly contributed to the higher sensitivity for the data quality. However, a number of challenges remain to be solved.

On the operational level, better quantitative metrics are required to measure higher semantic level quality aspects. CAD model reuse is particularly sensitive to hidden errors and anomalies. Design intent is still poorly addressed by MQT tools, but there are other bottlenecks for different stakeholders (Original Equipment Manufacturers (OEM), lower tier suppliers, and SME's) [40]. Profiles must become robust (changes do not produce unexpected failures) and flexible, e.g. modular (allow as many changes as necessary) [41]. If we can experimentally validate the hypothesis that the flexibility of the profiles depends not on the amount but on the semantic level of the constraints, valid metrics for flexibility may follow. For instance, detecting excessive use of poor 'fix' relations that lock point coordinates is an example of the type of high semantic quality tests that are not supported by current CAD check tools [32].

Dominant OEMs force top-down interoperability with their suppliers, which results in 'defensive' or 'conservative' designs, which are robust but hardly creative. Interoperability is a main concern for OEMs, as reusability is guaranteed by the best practices they impose, whereas simplification tasks are transferred to suppliers [42]. A hidden problem that hinders interoperability is the lack of proven modeling guidelines. Best practices are checked by MQT tools but also imposed and tuned by the OEM, whose current goal involves improving interoperability by abandoning explicit representations and adopting STEP AP 242. Validation rules require setup, and although quantitative metrics for shape errors already exist, they are context dependent and are governed by computational threshold values that are different for each MQT [41].

We see that the demands on data quality are getting higher and higher [42]. Therefore, the way practiced here appears to be very advantageous, preferably to use the models from a higher-quality library of CAD models. The alternative approach of performing object recognition via feature recognition in several stages requires a data quality check and subsequent repair procedure [43].

Considering a holistic approach, development of a data quality assessment tool—a dashboard—in addition to policies and protocols to manage data quality could be the solution for data quality issues on the enterprise level. Moreover, it should include systematic guidelines for planning the data quality assessment activity, extracting requirements for the data quality management, setting priorities to expedite the adaptation, identifying dimensions and metrics to ease the understanding, and visualizing these dimensions and metrics to assess the overall data quality [44].

Finally, a universal standard for describing, modeling, analyzing, measuring, testing, simulating, and building the real-world objects, products, and services remains a vision of Digital Twin. It promises to provide a platform whose standardized digital representation of real-world objects enables consistent, seamless exchange of technical information and interoperability across domains, industry silos, vertical markets, tools, and applications.

7.7 Conclusions and Outlook

As the preceding sections and use cases indicate, data quality plays an important role for the generation of Digital Twin. The generation of high-quality models is an important and integral part of digital processes today. The earlier approach of the 'digital master' has now evolved into a comprehensive approach of a Digital Twin. The developed approach with object recognition replaces the human hand in generation of Digital Twin providing the same level of data quality.

3D CAD models not only serve to depict the product shape geometrically, but also provide a basis for a large number of subsequent tasks and processes that are managed by the PDM system. The continuous further use and reuse of the CAD models by different users, such as CAE and CAM, can increase the effectiveness of virtual product creation and shorten the product creation time considerably [45]. As can be observed from the use cases, knowledge from downstream disciplines such as manufacturing must be implemented into data quality methods. At present, many of the original CAD models are either time-consuming post-processing by downstream users or, in some cases, even created from scratch. Many processes in PDM are still document-oriented [46]. Therefore, the sophisticated methods and techniques of quality management must be used for the virtual products of a digital planner as well as for physical products. Likewise, every model error leads to loss of time and additional costs. Ultimately, this knowledge has led to the use of fully-fledged parametric models to build the Digital Twin.

As a future development, a more comprehensive theoretical basis is required to define quantitative metrics for complex quality requirements (e.g. higher level semantics, data quality templates, relationship to Digital Thread). A new challenge for future developments represents the increasing product complexity caused by upcoming technologies, especially in area of electrics, electronics, and software [47]. As a typical supporting topic, data quality will keep the speed with the development of the subordinated systems.

References

1. Otto B, Österle H (2016) Corporate data quality. Springer, Heidelberg. https://doi.org/10.1007/978-3-662-46806-7
2. Haug A, Stentoft Arlbjørn J (2011) Barriers to master data quality. J Enterp Inf Manag 24:288–303. https://doi.org/10.1108/17410391111122862
3. Sha K, Zeadally S (2015) Data quality challenges in cyber-physical systems. J Data Inf Qual 6(2–3):8. https://doi.org/10.1145/2740965
4. Bondar S, Ruppert C, Stjepandić J (2014) Ensuring data quality beyond change management in virtual enterprise. Int J Agile Syst Manage 7(3/4):304–323. https://doi.org/10.1504/IJASM.2014.065346
5. Sebastian-Coleman L (2012) Measuring data quality for ongoing improvement: a data quality assessment framework. Elsevier, Amsterdam. https://doi.org/10.1016/C2011-0-07321-0
6. Sadiq S (2013) Handbook of data quality. Research and practice. Springer, Heidelberg. https://doi.org/10.1007/978-3-642-36257-6
7. Kraus M (2000) Entwicklung eines methodischen Konzeptes zur Sicherung der Datenqualität beim CAD-Datenaustausch, Diploma thesis, TU Darmstadt
8. Ofner MH (2013) Datenqualitätsmanagement aus Prozessperspektive: methoden und Modelle, PhD thesis, Universität St. Gallen
9. Yang J, Han S, Kang H, Kim J (2006) Product data quality assurance for e-manufacturing in the automotive industry. Int J Comput Integr Manuf 19(2):136–147. https://doi.org/10.1080/09511920500171261
10. Zhang R, Jayawardene V, Indulska M, Sadiq SW, Zhou X (2014) A data driven approach for discovering data quality requirements. In: Proceedings of the thirty fifth international conference on information systems: ICIS 2014
11. Pfouga A, Stjepandić J (2018) Leveraging 3D geometric knowledge in the product lifecycle based on industrial standards. J Comput Des Eng 5(1):54–67. https://doi.org/10.1016/j.jcde.2017.11.002
12. Schuh G, Rebentisch E, Riesener M, Ipers T, Tönnes C, Jank MH (2019) Data quality program management for digital shadows of products. Procedia CIRP 86:43–48. https://doi.org/10.1016/j.procir.2020.01.027
13. Boyd M (2009) Product data quality—different problem, different solutions. In: MIT information quality industry symposium, 15–17 July 2009. http://mitiq.mit.edu/IQIS/Documents/CDOIQS_200977/Papers/02_08_2D-2.pdf.

14. El-khoury J, Berezovskyi A, Nyberg A (2019) An industrial evaluation of data access techniques for the interoperability of engineering software tools. J Ind Inf Integr 15(2019):58–68. https://doi.org/10.1016/j.jii.2019.04.004
15. da Silva Serapião Leal G, Guédria W, Panetto H (2019) An ontology for interoperability assessment: a systemic approach. J Ind Inf Integr 16:100100. https://doi.org/10.1016/j.jii.2019.07.001
16. Wagner L (2020) Kein Digital Twin ohne digitale Durchgängigkeit. Zeitschrift für den wirtschaftlichen Fabrikbetrieb, Band 115, April 2020, pp 43–45. https://doi.org/10.3139/104.112315
17. Stark J (2015) Product lifecycle management, vol 1. Springer International Publishing, Switzerland. https://doi.org/10.1007/978-3-319-17440-2
18. Stark J (2019) Product lifecycle management (vol 4): the case studies. Springer Nature Switzerland AG. https://doi.org/10.1007/978-3-030-16134-7
19. Bondar S, Potjewijd L, Stjepandić J (2013) Globalized OEM and tier-1 processes at SKF. In: Stjepandic et al (eds) Concurrent engineering approaches for sustainable product development in a multi-disciplinary environment. Springer, London, pp 789–800. https://doi.org/10.1007/978-1-4471-4426-7_67
20. Biahmou A, Emmer C, Pfouga A, Stjepandić J (2016) Digital master as an enabler for industry 4.0. Adv Transdiscipl Eng 4:672–681. https://doi.org/10.3233/978-1-61499-703-0-672
21. Shen H, Bednarz T, Nguyen H, Feng F, Wyel T, Hoek PJ, Lo EHS (2019) Information visualisation methods and techniques: state-of-the-art and future directions. J Ind Inf Integr 16:100102. https://doi.org/10.1016/j.jii.2019.07.003
22. Emmer C, Fröhlich A, Stjepandić J (2013) Advanced engineering visualization with standardized 3D formats. In: IFIP advances in information and communication technology, vol 409. Springer, Berlin Heidelberg, pp 584–595. https://doi.org/10.1007/978-3-642-41501-2_58
23. Zilinski L, Sapp Nelson MR (2014) Thinking critically about data consumption: creating the data credibility checklist. In: ASIS&T proceedings of the American society for information science and technology, Seattle, WA, vol 51. https://doi.org/10.1002/meet.2014.14505101107
24. Young S, Palmer J (2007) Pedigree and confidence: issues in data credibility and reliability. In: 2007 10th international conference on information fusion. https://doi.org/10.1109/ICIF.2007.4408053
25. Belkoura S, Zanin M, LaTorre A (2019) Fostering interpretability of data mining models through data perturbation. Expert Syst Appl 137:191–201. https://doi.org/10.1016/j.eswa.2019.07.001
26. ISO 8000-60 (2017) Data quality—Part 60: data quality management: overview
27. Lee C, Kim S, Lee J, Lee D, Caballero I (2019) Perez del Castillo R, Yan H (2019) Data quality assessment for manufacturing processes. ICIC Express Lett Part B: Appl 10(12):1107–1113. https://doi.org/10.24507/icicelb.10.12.1107
28. ISO 8000-61 (2016) Data quality—Part 61: data quality management: process reference model.
29. ISO 8000-62 (2018) Data quality—Part 62: process data quality management: organizational process maturity assessment. Appl ISO/IEC 330xx Family Stand
30. IEC 62264-1 (2013) Enterprise-control system integration—Part 1: models and terminology
31. Fischer T, Martin HP, Endres M, Stjepandić J, Trinkhaus O (2000) Anwendungsorientierte optimierung des neutralen CAD-Datenaustausches mit Schwerpunkt Genauigkeit und Toleranz, VDA, Frankfurt
32. N.N., 4955 (2006) Umfang und Qualität von CAD/CAM-Daten, VDA, Frankfurt
33. N.N., ISO/PAS 26183:2006 (2005) SASIG-product data quality guidelines for the global automotive industry. International Organization for Standardization, Switzerland
34. Kuhn O, Liese H, Stjepandić J (2011) Methodology for knowledge-based engineering template update. In: Cavallucci D, Guio R, Cascini G (eds) Building innovation pipelines through computer-aided innovation, Springer, Berlin Heidelberg, pp 178–191. https://doi.org/10.1007/978-3-642-22182-8_14
35. Paukkunen L (2018) Development of metrics and automation for product model verification, Master thesis, Aalto University

36. González Lluch MC, Company P, Contero González MR, Camba JD, Colom J (2017) A case study on the use of model quality testing tools for the assessment of MCAD models and drawings. Int J Eng Educ 33(5):1643–1653
37. Denkena B, Dittrich MA, Stobrawa S, Stjepandić J (2020) Efficient generation of a digital twin using object detection for data acquisition and XML-interface for model creation. Procedia CIRP 93:274–279. https://doi.org/10.1016/j.procir.2020.03.016
38. Yang W, Fu C, Yan X, Chen Z (2020) A knowledge-based system for quality analysis in model-based design. J Intell Manuf 24. https://doi.org/10.1007/s10845-020-01535-8
39. Hirz M, Rossbacher P, Gulanová J (2017) Future trends in CAD–from the perspective of automotive industry. Comput-Aided Des Appl 1–8. https://doi.org/10.1080/16864360.2017.1287675
40. González-Lluch C, Company P, Contero M, Camba JD, Plumed R (2017) A survey on 3D CAD model quality assurance and testing tools. Comput Aided Des 83:64–79. https://doi.org/10.1016/j.cad.2016.10.003
41. Ostrosi E, Stjepandić J, Fukuda S, Kurth M (2014) Modularity: new trends for product platform strategy support in concurrent engineering. Adv Transdiscipl Eng 1:414–423. https://doi.org/10.3233/978-1-61499-440-4-414
42. Sun J, Hiekata K, Yamato H, Nakagaki N, Sugawara A (2014) Virtualization and automation of curved shell plates manufacturing plan design process for knowledge elicitation. Int J Agile Syst Manage 7(3/4):282–303. https://doi.org/10.1504/IJASM.2014.065348
43. Bondar S, Shammaa A, Stjepandić J, Tashiro K (2015) Advances in parameterized CAD feature translation. Adv Transdiscipl Eng 2:615–624. https://doi.org/10.3233/978-1-61499-544-9-615
44. Gürdür D, El-khoury J, Nyberg M (2019) Methodology for linked enterprise data quality assessment through information visualizations. J Ind Inf Integr 15:191–200. https://doi.org/10.1016/j.jii.2018.11.002
45. Stekolschik A (2007) Ein Beitrag zum ganzheitlichen Qualitätsmanagement von CAD-Modellen in der Produktentstehung, Dissertation, Ruhr-Universität Bochum
46. Lämmer L, Theiss M (2015) Product lifecycle management. In: Stjepandic J, Wognum N, Verhagen WJC (eds) Concurrent engineering in the 21st century. Springer International Switzerland, pp 455–490. https://doi.org/10.1007/978-3-319-13776-6_16
47. Van Der Velden C, Bil C, Xu X (2012) Adaptable methodology for automation application development. Adv Eng Inform 26(2):231–250. https://doi.org/10.1016/j.aei.2012.02.007

Chapter 8
Object Recognition Findings in a Built Environment

Josip Stjepandić, Sergej Bondar, and Wjatscheslaw Korol

8.1 Introduction

Like other assets of a manufacturing company, the modern factory is also planned and documented by using powerful software tools which generate high-fidelity factory models. A virtual factory layout is the main digital representation of a factory ("Digital Master") which includes all necessary constituents and equipment (building, machinery, and auxiliary equipment) [1]. In the focus of the layout planning are the design and the maintenance of the production system which determines the productivity and feasibility of the entire factory [2]. There are also a number of principles and guidelines for how to design the value stream in the factory for several criteria, e.g. short lead-times or high efficiency [3]. To achieve an optimal planning result, it is necessary to share ideas between several teams or employees and functions which can be made by using collaboration front-end of manufacturing systems [4]. Virtual factory layout is beneficial for communication and knowledge transfer since different people from all around the world can take part of information without being present [5].

In general, planning systems need to facilitate the evaluation of the targeted factory design feasibility. By powerful 3D CAD systems, designers are being enabled to estimate the actual setting of that factory design [3]. CAD system provides the functionality to build models and templates which can drastically automate some tasks (e.g. setup, analysis, or repeating tasks) [6]. Small changes and complete re-layout

J. Stjepandić (✉) · S. Bondar · W. Korol
PROSTEP AG, Dolivostr. 11, 64293 Darmstadt, Germany
e-mail: josip.stjepandic@prostep.com

S. Bondar
e-mail: sergej.bondar@prostep.com

W. Korol
e-mail: wjatscheslaw.korol@prostep.com

J. Stjepandić et al. (eds.), *DigiTwin: An Approach for Production Process Optimization in a Built Environment*, Springer Series in Advanced Manufacturing,
https://doi.org/10.1007/978-3-030-77539-1_8

could also be easily accomplished. Inside the 3D environment, the depth perspective and better interactions could be achieved which is not possible in a 2D view. Potential problems such as safety issues, aisle and other layout problems can be visualized and modified by applying plant layout problem solving techniques. Re-layout of the existing factory layout can be done until satisfactory result e.g. material handling system is obtained [5].

There are several prerequisites and criteria to consider when planning layouts, which makes it suitable to use virtual tools. A common type of virtual tools used is layout modeling using 2D CAD applications, with data of the existing shop floor area from previous layouts or blueprints. There are problems found with working only with such virtual tools for the layout planning. One such problem is that the data can contain errors originating from measurement mistakes or undocumented changes in the production systems. Another problem is to communicate and discuss the planned layouts, which can lead to misunderstandings between employees. These problems can result in errors that are causing problems when implementing the planned layout [7]. If 3D CAD systems are used, the consistency and accuracy of the resulting models are improved, as well as spatial operations can be modelled and processed.

During progress of the construction process, significant differences between the construction states "as-planned" and "as-built" can occur. The differences between planned and real implemented layouts can lead to difficulties when building a valid simulation model due to a lack of proper and accurate spatial data. Hence, the simulation model might be missing many details without being noticed due to being abstract. This abstract simulation model can yield a simulated capacity that is impossible to achieve when the real production system is running due to a lack of details and information in the model, and investments not giving the expected results [7].

Another application of virtual factory layout is to provide time and cost savings by supporting offline layout design and simulation of factory operations in a Digital (Manufacturing) Twin as described in Sect. 3.2.2. Evaluating design changes and process improvements offline increases time for planning and decreases time for realization, which enables the real production to not be disturbed as much as if improvements were evaluated in the real existing production system. Further, a virtual factory layout enables execution of product and process development and improvement in parallel [1].

There are several advantages of using virtual factory layouts for the improvement work of an existing environment. High transparency is an overall benefit for all related business processes. However, while creating a virtual factory layout, it is important to prevent design mistakes as well as measurement and modelling inaccuracies due to the cost it would cause. All objects in the factory need to be positioned in an accurate digital environment [8]. 3D laser scanning is an accurate measurement technology which captures as-built data that can be used as reference for accurate modelling as well as for dimensional check [1].

There is a need to ensure the spatial data of the shop floor area used for layout planning are geometrical accurate and up-to-date. 3D laser scanning has been shown to be a fast and efficient technology for capturing such spatial data. The 3D laser scanning results in point clouds that can be used for layout planning by combining them with 3D CAD models of the new equipment into realistic 3D layout models as presented in Chaps. 4 and 6. These layout models can be communicated and discussed during the layout planning process with different stakeholders of the layout. The main challenges are to work efficiently and systematically with such models during the planning process and to ensure all parts of the production system are included. A key part of this work is to create a common factory model for all stakeholders, which should include the virtual model and the reality [3].

As presented in the previous chapters, point clouds can be used as a basic geometrical description for a factory in term of reverse engineering [9, 10]. However, the application of the point clouds is not exhausted with this [11]. From the point of view of the construction, point clouds are adopted in various applications throughout construction project lifecycle from planning and design phase to fabrication and construction phase, and to operation and maintenance phase [12]. The applications and benefits of point cloud data in different project phases are summarized in Fig. 8.1 which can be understood as constituents of Digital Twin. In the planning and design phase, point cloud data are used to collect the geometric information of the construction sites and existing buildings to assist in project planning and design [13]. The project quality and productivity are expected to improve due to the better decision-makings with accurate site data.

In the fabrication and construction phase, point cloud data are leveraged to check the geometry quality of precast and prefabricated construction components as well as completed construction works on site. Construction progress can also be tracked using point cloud data in a more efficient and automated manner. Construction safety is enhanced by using point cloud data for the identification of safety hazards, identification of blind spots of construction equipment, assisting in crane operations, as well

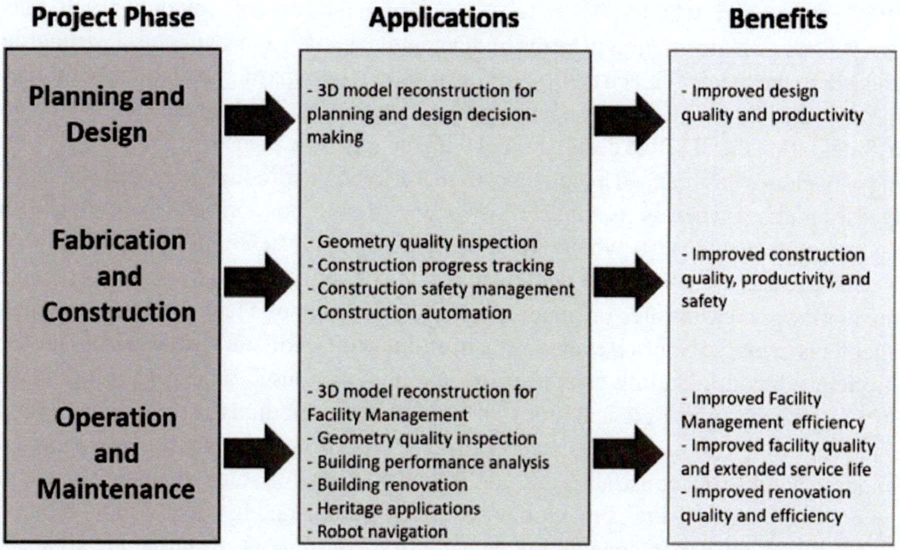

Fig. 8.1 Usage of cloud point data in different project phases [11]

as training of site workers. Furthermore, point cloud data are also useful for digital reproduction of construction components as well as automated earth excavation system. In conclusion, point cloud data have been utilized to improve the construction quality, productivity, and safety, by providing accurate geometric and semantic data for construction objects [11]. Thus, the use of feature CAD models is beneficial, if the corresponding objects can be recognized in the point cloud accurately.

In the operation and maintenance phase, by creating 3D models of the existing facilities from point cloud data, facility management can be done in virtual 3D environments with improved efficiency. The geometry quality of facilities is checked routinely to keep smooth operation and timely repair of any defect. In addition, point cloud data are adopted for building performance analysis in various aspects so that the facility managers have a better idea on the building performance [11]. Specific audits such as creating of the inventory directory (e.g. identification and localization of fire extinguishers and fire alarms) can be provided by fast scan of the rooms by a mobile device and subsequent recognition of objects of interest [14].

Building renovation, retrofit, and refurbishment also benefit from point cloud data because of the accurate and efficient modelling of the existing buildings [15]. Similarly, heritage preservation, repair, and maintenance use point cloud data due to the geometric and semantic information that could be extracted from point cloud data. Robot navigation is also replying on point cloud data, which eventually improve the operation and maintenance quality of a facility. Consequently, point cloud data are expected to bring improved facility management efficiency, improved facility quality and extended service life, as well as improved renovation quality and efficiency in the operation and maintenance phase [11].

Finally, it can be said that point clouds are a widespread type of data acquisition in indoor and outdoor applications, also adopted within Building Information Modelling (BIM) technology, which is becoming a standard in the construction industry. BIM is a rich source of information in terms of 3D geometry, such as architectural, structural and MEP (mechanical/electrical/plumbing), knowledge sharing, and interoperability through the whole lifecycle of a building. A so-called 4D BIM model can be created by integrating the 3D BIM design model with the schedule plan [16]. Data acquisition of point clouds has reached a high level of maturity in the productive use within BIM. Question arises when is reasonable to work with pure point clouds and when with feature models of objects which are to be recognized in such clouds.

Since it was not possible to put this multitude of application options into practice, the work was concentrated on generation of Digital Twin of a factory. The aim of this chapter is to investigate and evaluate the usefulness of a realistic virtual factory model in factory layout planning primarily for Digital Twin based on object recognition. It was realized by a practical study of how existing methods for data acquisition and processing can be concatenated in a robust workflow and, subsequently, applied under real industrial constraints and conditions. During this study, realistic 3D layout models are created using point clouds acquired as described in Chap. 4 and prepared as described in Chap. 6 considering data quality requirements described in Chap. 7. This workflow was set up and its results were evaluated in industrial workshops. The participants from the user side were engineers involved in the factory layout planning and machine operators that will work within the production system.

The structure of this chapter reflects these aims. In the following Sect. 8.2, the definition of process design and its supporting concept are briefly introduced. Automation of workflow with respect to different data flows and conflicting requirements is discussed in Sect. 8.3. Subsequently, procedure of the practical model preparation for object recognition process is drawn in Sect. 8.4. Section 8.5 showcases the model segmentation in practice. Training and testing of objects in a factory are presented in Sect. 8.6. A discussion Sect. 8.7 gives insight into benefits and gaps of the presented approach as well as future directives. Finally, an outlook is given in Sect. 8.8 with respect to the future improvements and importance of object recognition from industrial process perspective.

8.2 Process Design

A framework must be defined, based on a commercial software system, to facilitate the biunivocal relation between a physical factory or a construction object and its equivalent digital counterpart which can be integrated with production system and BIM. Although this framework primarily translates the point cloud model in an object representation, it considers the different topics and collects the necessary information involved in the creation of a production system digital counterpart, i.e. complexity, identification, lifecycle, information and configuration, and the main software applications involved. Hence, the implications on the digital counterpart creation from the view of the industrial design should be considered too [17].

As described in Chap. 3, Digital Twin has a few pre-requisites [18]. First of all, it is the availability of the high-quality models of all involved objects [19, 20]. In Sect. 6.6.4 (in particular illustrated in Fig. 6.10), four methods for creation such models are shown. The first two paths reflect the recognition of objects, and the next two the creation of models based on recognized singular features of an object. While these paths can run almost simultaneously and independently, a comprehensive workflow control and management are necessary to preserve a stable and reliable process [21]. The most complex path belongs the object recognition path which requires a complete installation of a standard machine learning framework (Theano, TensorFlow or Caffe, according to Table 6.4) with complementary utilities. Finally, the importance of data quality is highlighted in previous Chap. 7, based on previous research [22].

Depending on the selected object recognition method (Table 6.4), the model transformation must be implemented in an appropriate way which can comprise additional translational steps. VoxNet and VoxelNet, for example, transform the point cloud into voxels (3D grid cells) that each contain a small amount of points [23]. It produces bounding boxes based on features of the voxels. G3DNet, as a point-based method, is a DL architecture that is used in 3D object classification and segmentation. It attempts to semantically segment each point in the data by learning the local and global features of the points and classifying each of them [23]. Clusters of points with the same labels can be detected as objects. Nevertheless, usage of voxels has the side effect of simplification which must be considered.

The previous research has shown that the differences in accuracy of selected methods (VoxNet, VoxelNet, PointNet) are significantly lower than evaluated in the ModelNet benchmark [24, 25]. This low scattering of results can be attributed to the fact that these are predominantly prismatic objects that differ more in terms of their local rather than global features. First of all, this places considerable demands on the level of detail of the sample models for training [25]. Secondly, this finding makes the selection much easier. Nevertheless, the selected workflow must be flexible and provide the possibility to embed object recognition methods which use different algorithms. The main criteria for selection of the object recognition framework are listed in Table 8.1.

The criteria with a high and medium weight are marked with ++ respectively + signs. High accuracy, augmentation of data and robustness are essential to create a stable and reliable workflow [26]. The complexity and availability of the framework (criteria low parameter number, integrated segmentation, open-source availability) have a significant impact on expenses rather than quality of results. Lower parameter number indicates shorter processing time bot for training and testing as well as lower hardware requirement. While multiple dozens of approaches with similar accuracy exist and are being developed further, additional modules such as the integrated segmentation (e.g. PointNet [24]) helps in further modularization [27].

Apart of costs, the availability of the stable, open-access framework proven by a large community at GitHub or similar platforms promises a continuous maintenance for a certain period with rapid bug-fixing. Therefore, assuming a flexible workflow structure one of these frameworks is preferred. The criteria real-time execution and low resource demand indicate the demand that the executable software runs on standard and mobile hardware (tablet or smartphone). Considering the complexity of the entire process, this requirement currently looks unattainable [28]. However, the question remains whether the singular steps can be executed on mobile devices.

Table 8.1 Criteria for selection of the object recognition framework

Criterion	Weight/degree of fullfilment
Accuracy	++
Augmentation of data	++
Robustness	++
Low parameter number	+
Modular structure	+
Integrated segmentation	+
Open source availability	+
Broad installation base	+
Real-rime execution	0
Low resource demand	0

Obviously, there are many conflicts of goals and the selection must be done using the best compromise. An implementation of two or three frameworks in parallel could be a pragmatic solution to resolve outliers. Flexibility is further necessary on augmentation and building of secondary models. For training, a basic number of relevant objects should be selected and transformed into point cloud using a virtual scanner. Training models can also be extracted from existing point clouds. Augmentation of data should be done with 24 rotations. If voxels are used, voxels with either fix or variable size can be selected.

8.3 Automation of Workflow

Complex procedures as described in Sect. 6.6 usually are being controlled by appropriate workflow software. Prior of such an implementation, it must be defined how the process steps scan, object recognition and generation of Digital Twin (Fig. 3.7) interact with each other. Basically, that are tasks for different user profiles because the present technology does not provide the performance to collect entire functionality for generation of Digital Twin in one device (e.g. portable computer). The tasks scan, object recognition, design engineering (to provide the requested change) and process planning (generation of Digital Twin) can be seen functionally independently of each other; with data exchange among singular steps [28].

Complex workflow which includes data exchange with partners requires some additional topics to be addressed [26]:

- Security: Public networks are open to everybody; sensitive information needs to be exchanged in a secure way.
- Reliability: Public networks are often not as stable as required, especially for transmission of large amounts of information (e.g. point cloud model file pack-age which contains dozens of gigabytes).
- Traceability: For the exchange in a globalized economic environment often—even legally binding—prove of execution, data transmission and reception is required.
- Efficiency: Process robustness and stability (such as repeatable exchanges and defined content) without loss of competitiveness becomes more and more crucial.

A deployment schema for generation of Digital Twin is depicted in Fig. 8.2 as a collaborative procedure. The scan is conducted on the customer's premise. Then point cloud data are sent to OpenDESC.com where object recognition is done as described in Sect. 6.6. Results are transmitted to the process planning where the Digital Twin is finally built.

Assuming a request for reconstruction, overhaul, or maintenance, the object acquisition must be provided at the production site by any device (scanner or camera, fix or mobile) and the raw data are transferred by a secure web connection or disk to the next instance which provides data processing (object recognition, feature recognition, translation) as described in Sect. 6.6. After the full data model of the factory was created, it is sent to the design and engineering which complete the required change of

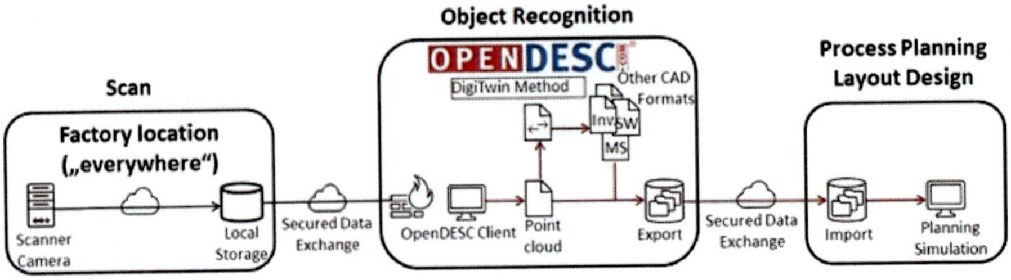

Fig. 8.2 Collaborative generation of Digital Twin [26]

the production system. Finally, the data set is being exported to the process planning department [21].

While some steps must be done manually (point cloud preparation, noise filtering, outlier removal, feature extraction, and almost the entire training), the workflow has to ensure an automated execution of all remaining steps. This is in particular true for the most time-consuming steps: execution of segmentation and execution of object recognition. It also includes the handling of alternative routes and loops which are necessary to execute all the options illustrated in Fig. 6.10. The flexible, modular handling of the object recognition framework with the option of replacement by an another must also be provided [25].

Further important aspect is the level of automation of singular steps as illustrated in Fig. 8.3. While singular data sets differ from each other, for sake of efficiency is it necessary to estimate the share of manual assistance in the entire process. The device-oriented tasks already are highly automated and can provide a dense point cloud as input for a high-performance point-cloud editor in particular if a scanner is used.

The most remaining manual work can be expected on the segmentation of complex scenes which promises the highest rationalization potential, although no support from

Steps and sub-steps	Automation level	Steps and sub-steps	Automation level
Feature extraction	+	Registration	+
Feature matching	+	Noise filtering	~
Camera motion estimation	+	Outlier removal	~
Sparse 3D reconstruction	+	Down-sampling	+
Model parameters correction	+	Mesh reconstruction	+
Dense 3D reconstruction	+	Point cloud segmentation	~
Binocular camera calibration	~	Point cloud modeling	+
Absolute scale recovery	~	-	-

Note: (+) achieved, (~) partially achieved.

Fig. 8.3 The automation level of singular steps in 3D reconstruction techniques [11]

publicly available directories can be expected. A successful segmentation anticipated a reduction in data volume of point clouds with a factor of 1.000 and more between the rough point cloud and the segmented point cloud focused on segment for further processing can be achieved. This drastically reduces the demand on hardware, the processing time and makes the interactive work much easier [18].

8.4 Model Preparation

Each model must be processed according to the procedure described in Sect. 6.6. In particular, the point cloud must be prepared for the object recognition. At first, the sighting must be premised to visually check the output of the data acquisition procedure. It is possible that the automatic registration has failed. Therefore, the data set must be positioned in space. It occurs by selection of an appropriate reference coordinating system. During these manual steps, the model clean-up, the noise filtering and the outlier removal are conducted. Eventually, the entire point cloud must be splitted, e.g. to extract the portion of interest which will be processed further. The extraction of models and portions for training of recognition respectively segmentation occurs at this point of time. Finally, the point cloud must be exported in the desired format (e.g. e57).

From this point on, the process runs automatically apart of training activities which are not mandatory for each data set. Training is necessary in a sufficient extent to provide the test base for the later testing [29]. Mesh reconstruction and point cloud segmentation are implemented as separate utilities for the object recognition framework. It is necessary for voxel- or surface-oriented frameworks. To collect all automatic running steps, a user front-end was built (Fig. 8.4). It controls the semantic segmentation, cluster extraction and cluster classification (recognition of singular objects). This tool includes also a viewer which allows the visualization of each intermediate result.

A point cloud model of an entire factory is a huge model of several dozens of Gigabyte memory space which is challenging and time-consuming from the basic data handling on. The preparation steps have also the aim to reduce the volume of data to an extent which is much easier for the interactive work. Besides, it is beneficial to split it into multiple sections according to the structure and layout of the factory. Afterwards, the singular pieces can be considered as component of an assembly. It makes the work easier as well as it simplifies the distributed work and use of the parallel algorithms.

Using afore-mentioned methods and utilities, a reduction of data volume of the point cloud by a factor up to 1000 becomes necessary without losing the quality of the content. The range of volume reduction heavily depends on the desired object recognition framework.

Fig. 8.4 The user interface of the DigiTwin tool

8.5 Model Segmentation

Although good object recognition algorithms are capable to search for objects in each space, for an intended industrial exploitation some pre-processing and adjustment in sense of subdivision (segmentation) of huge spaces (and data volumes vice versa) look promising. The segmentation of a scene is beneficial for several practical reasons. At first, it helps to generate a structure of a scene which aims to identify specific points of interest. Secondly, it can help to extract unnecessary data (remaining outliers, objects which were scanned by a mistake) in a structured manner. Thirdly, the segmentation helps facilitates adding of further information (e.g. buffer, hold places, restricted areas). Fourthly, it is much easier to validate the object recognition results and handle the failures and errors in a restricted, smaller space. Finally, a segmented scene allows the use of different recognition methods for different classes of objects (e.g. machines vs. pipelines).

Segmentation can be conducted manually or by supervised training which is the preferred way for complex scenes. Therefore, additional methods were developed to simplify object recognition preserving the data quality. The first issue is the preselection of smaller agglomerations in the entire point cloud which are assumed to contain an object, by using a PCL module. The initiative PCL (Point Cloud Library) presents an advanced and extensive approach to the subject of 3D perception of point clouds, and it is meant to provide support for all the common 3D building blocks that applications require [30]. The library contains state-of-the art algorithms for: filtering, feature estimation, surface reconstruction, registration, model fitting and segmentation. PCL is supported by an international community of robotics and perception researchers. In particular, the functions for filtering and segmentation of a point cloud are used in the presented approach. The segmentation module is being trained to subdivide the entire scene in nine categories: roof, wall, floor, machines, structure, cranes, transport, auxiliary, and noise. For a mechanical shopfloor, machines, cranes, and transport devices are of a particular interest.

Especially, the segmentation training consists of the following steps:

1. For all halls (rooms), extract different semantic classes as a single region using Recap
2. Extract regions as *ply* files
3. Train PointNet using all the halls
4. Input: point cloud of a hall as a scene
5. Divide the scenes in blocks
6. Block size: 1, 2 m
7. Simple number of points: 4090, 1000, 12,000
8. Train using different blocks
9. Merge all blocks together
10. Output: point cloud of the hall assigned with a semantic class.

As depicted in Fig. 8.5 which shows the segmentation results of a hall with nine object categories in three rows, cluster extraction in the point cloud of a hall is executed in three consecutive steps. At first, the point cloud must be reduced (sparsed) by using a voxel grid filter without losing the content. Subsequently, some unnecessary objects (floor, walls, and the roof points) are removed by using the normal segmenter. It is worth to note that the extent of categories and objects to be filtered out is determined by the final utilization of the point cloud. For material flow planning floor, walls, roof, and thin pipelines are not necessary. Finally, all existing clusters in

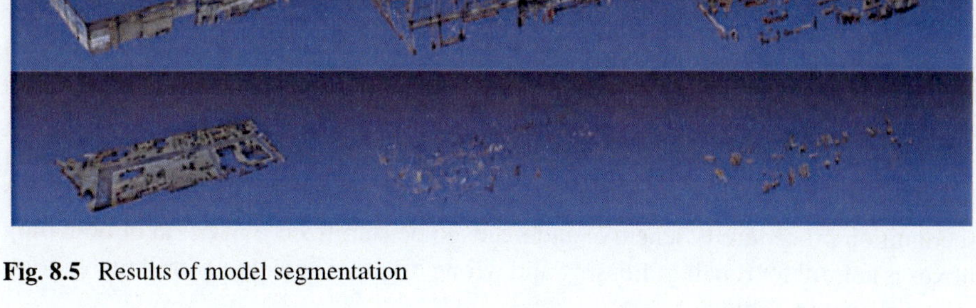

Fig. 8.5 Results of model segmentation

the hall are extracted by using the Euclidean cluster extraction. The outcome of this module consists of a crowd of different clusters which subsequently are being used for object prediction: which objects does contain the corresponding point cluster. That is the object recognition in the tight sense: testing of pretrained CNN. Furthermore, for each cluster, a bounding box is extracted which is oriented in main axes of the point cloud. Such a bounding box is the first aid for visual control whether an error has happened and replaces a missing object.

8.6 Training and Testing, Rebuild of Structures

Before the object recognition can start, the objects of interest must be available in the system library. Since they are usually not included in the standard database such as ModelNet (Sect. 6.4), an additional reference library must be set up that contains the objects of interest. This is performed in two ways: either by using CAD models or exemplary point cloud models. To ensure the similar properties of both model types, CAD models are transformed in point clouds using a virtual scanner (Fig. 8.6) which is stochastically moved around the object (illustrated with red rectangle and blue points) in five runs with variable distances. CAD models can either be provided by the object manufacturer respectively vendor or have to be built up manually first. The similar objects can be derived from the existing properties with little effort.

Singular point clouds models must be further transformed in order to prepare an adequate input for object recognition procedures, depending on the object recognition framework used [31]. In case of the voxel-oriented framework (VoxNet, VoxelNet), the point cloud must be translated into a corresponding voxel model. The side length of a voxel plays a significant role here. Tests have shown that a variable length which is related to the object shape provides the best results. Figure 8.7 illustrates the translation of a CAD model of a machine to a point cloud model and, finally, a voxel model. Although the start model already has been simplified, a further simplification occurs during the translation to a voxel model which hardly can be prevented. If the simplification dilutes a local feature (e.g. an auxiliary device), then sufficient recognition results cannot be expected. Rather, it is to be feared that a wrong object will be recognized. Therefore, when building the sample models, care must be taken to ensure that the distinctive features (e.g. oil purificator) are considered.

An exemplary procedure in three steps (segmentation, clustering, recognition of object in singular clusters) is illustrated in Figs. 8.8, 8.9 and 8.10 [28]. After the sample objects (machines, cranes, trolleys) have been trained with up to 1000 epochs, the developed framework has been tested using real-life data extracted from the factory hall. At this time, the particular importance was given to the object recognition. The overall recognition rate is high, although the environmental impact is quite negative. More than 75% of all objects can be truly recognized. However, the recognition procedure is sensitive and needs to be stabilized. Provision of bounding boxes is helpful for repair of missing and wrong oriented objects. Overall recognition rate is shown in Table 8.2:

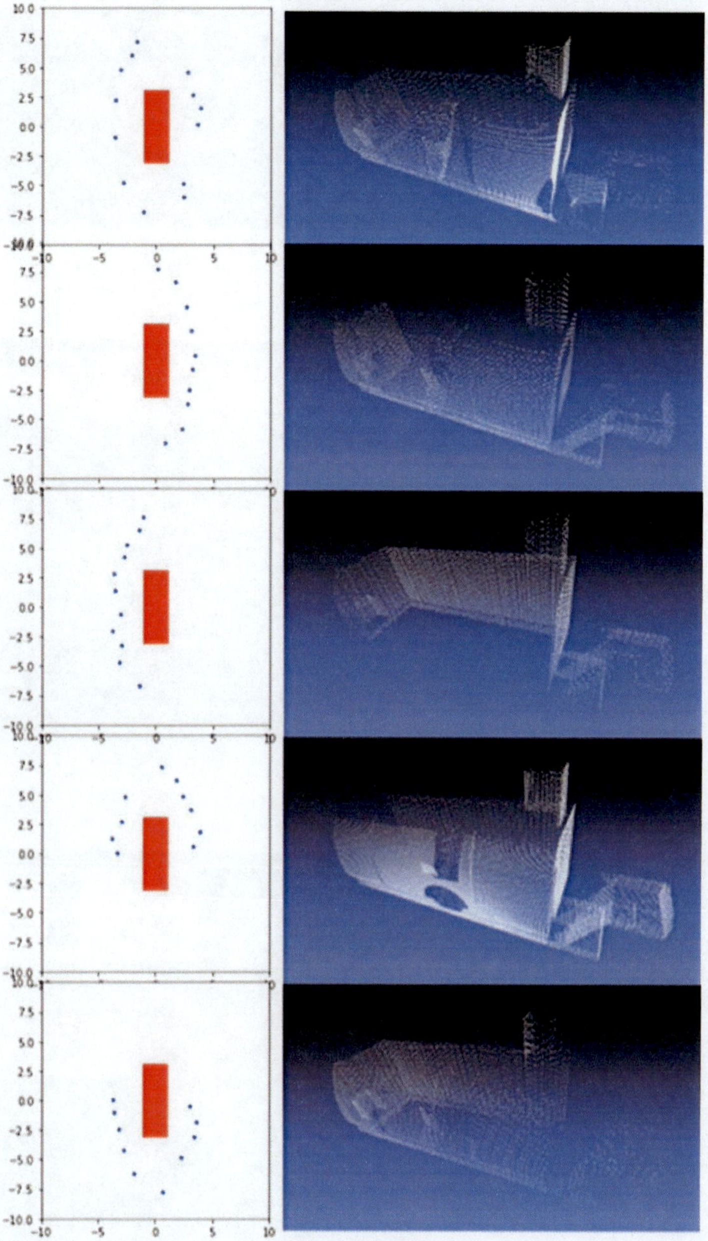

Fig. 8.6 Generation of a point cloud by a virtual scanner

Fig. 8.7 Generation of the voxel model of a machine

Fig. 8.8 Factory hall as a point cloud scene

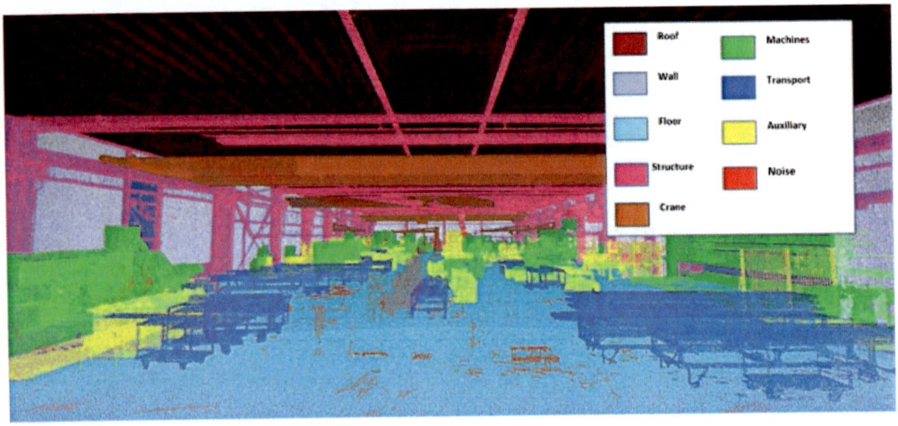

Fig. 8.9 Factory hall as a point cloud scene after the segmentation in 9 categories

Figure 8.8 shows the screenshot of the factory hall as a scanned scene. Apart of the point cloud, modern terrestrial laser scanner additionally generates high-resolution images of the scene which can be used for additional analysis and documentation [32] (Chap. 4).

Fig. 8.10 Factory hall as a scene with recognized objects

Table 8.2 Inventory of the factory (excerpt)

Objecty type	Recognition rate %
Machine	80
Intralogistics	80
Storage, container	80
Pipeline	80
Furniture	70
Mobile object	70

Figure 8.9 shows the screenshot of the factory hall as a scanned scene after the segmentation conducted in nine categories which are marked by different colours each. It is easy to identify that all points are assigned to an auxiliary category, except for small clusters on the floor that can be seen as noise (category 9). Such structure is important for the following steps: recognition of singular objects and material flow simulation.

Figure 8.10 shows the screenshot of the factory hall as a scene built with recognized objects which are put in the right position and orientation in space. For the better distinction, different objects are marked with different colours. Such a structure is being exported to CAD for repair of remaining errors and completion by high fidelity CAD models as input for subsequent tasks (e.g. simulation of material flow).

On the one hand the rule of thumb applies that all objects with a unique, distinctive shape like a hanging crane can be recognized easily in the right position and orientation in space (Fig. 8.11). On the other hand, there are objects that have a similar shape, without local features, and can therefore easily be confused, even if they have very different functions (container, housing, tank). In such a case, utilization of a secondary criterion (e.g. labeling) for the reliable object recognition could be also considered. During the analysis of the outliers (objects which were either not recognized or recognized false) three main drawbacks became apparent which cause the failure of recognition: occlusion, small test base and overfitting. Of course, there is a significant, but not full inference of the overfitting and the small test base.

Fig. 8.11 Recognized object: hanging crane

Otherwise, the extension of the test base which happens continuously by adopting additional models will provide a better training and, subsequently, better results.

An algorithm which considers only a single part in a point cloud cluster fails completely in case of a significant occlusion; when two or more objects next to each other are assumed as one cluster (Fig. 8.12). Although this case does not rarely occur

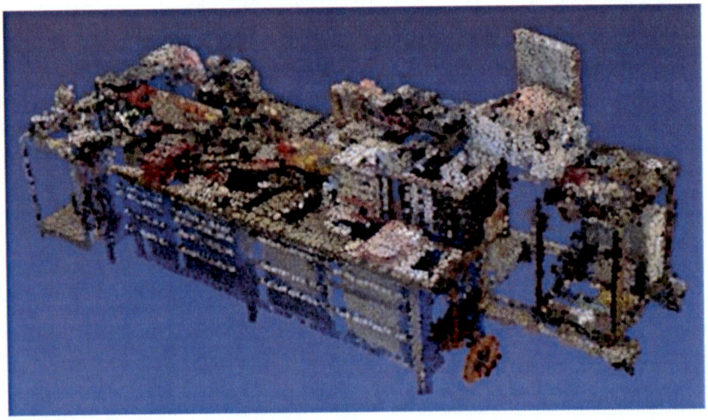

Fig. 8.12 Impact of occlusion

in a factory (Sect. 6.8), it will not be investigated further [33], but resolved by an additional loop, where clusters which contain more than one object are subdivided manually and then processed separately. Just worth to remark that a fully cluttered hall would stop the whole undertaking of generation a Digital Twin by means of object recognition.

Scanning with a scanner device in the height of approximately one meter above the ground of the factory has a basic drawback that the top area (the roof) of the large objects cannot be acquired sufficiently. A similar case arises when using a hand scanner. That can be resolved, for example, by using a drone as the scanner carrier [34]. In order to create a closed geometrical object which can be automatically translated to a voxel representation, the top of the object is being approximated by a plane in order to close the volume. That is not only a dimensional deviation, but also reduces the possible distinctive characteristics of the object [35]. This drawback is enforced by a basic structural difference between the test object which is derived from an exact CAD model by using a virtual scanner and the scanned object.

Overfitting is rarely obvious, but its impact is ubiquitous. All three used CNN frameworks (VoxNet, VoxelNet, G3DNet) provide similar results and are sensitive on change (e.g. input data quality). This implies a strong overfitting. The attempts to reduce overfitting lie in the extension of test base by more model variants. Just as important is the compromise between the level of detail, which takes local features into account, and a limited model complexity for each object in the object library. Such considerations should also consider the modular structure of the object models, similar to how the machines are built [27]. This can be provided by parametric structure of sample CAD models.

Process-compatible CAD models are required for further processing of the data in the simulation of the material flow [36]. For this purpose, a multi-stage assembly is created in the CAD system that depicts the structure of the factory hall (Fig. 8.13).

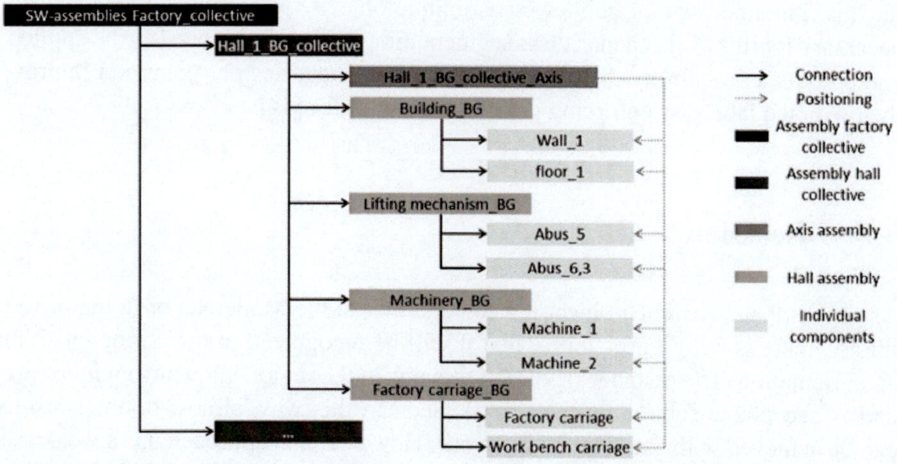

Fig. 8.13 Concept structure for a template for factory hall

For each object, a CAD model from the library is inserted for each object found. At the same time, a bounding box is inserted that describes the exact position in space. The objects found are summarized in a list that describes an assembly group in the hall which can be imported in a CAD system [37].

The assembly in the CAD system SolidWorks as shown in Fig. 8.13 represents an industry standard in the field of industrial engineering and planning. The whole hall is divided into several cells, each containing machines and devices, which can then be further subdivided as required. Apart of the training process which lasts two days on a desktop computer, this automated process takes few minutes of computing time with potential for further improvement. For further work (e.g. customization design) a complete, consistent CAD model is necessary, which is validated by a quality check [38, 39]. If some components are missing here, they must be added later. Missing geometry can be created and adapted by using standard CAD functions (Sect. 6.6.4). The bounding box is of great advantage here, because the missing component is placed in it. Once the assembly has been completed, it can either be used in a customization design or exported to STEP or XML for simulation tasks [40]. Ultimately, this is how Digital Manufacturing Twin is created in a built environment [41].

In order to verify this approach for the recognition of pipelines, further experiments with the point cloud from the plant construction (biogas plant) have been conducted. This method is only based on the registered industrial point cloud itself regardless of the varying point densities [42]. Such industrial point clouds contain defects such as occlusions and sparseness. The results were disparate. Good results were achieved in the segmentation phase automatically and robustly. Pipeline easily can be separated from the remaining components of a plant. In highly dense industrial spaces, class segmentation of cylinders, I-beams and valves that have easily distinguishable geometric patterns in the processing unit, warehouse and the oil refinery work well. Object recognition does not perform sufficient. There are several reasons for this. Pipelines cannot be viewed as unique objects. Alternative approaches are necessary for this [43]. Then, a class segmentation deep network needs to be applied, and outputs a probability distribution of all label categories per point and improve the predicted labels by enforcing post-processing rules [44].

8.7 Discussion

Assuming that a searched object is known, either in the ModelNet or in the private library, there is a high probability that it will be recognized in the correct position and orientation. The distance deviation remains in the region known from literature, and is acceptable for this purpose [45]. Seen in this way, almost optimal results can be achieved with small manual rework. However, this approach has a weakness due to the small test base for industrial products. Here, verification in larger space with more complicated scenes and different types of repetitive objects is needed. A collection like ModelNet for industrial objects in a factory would be supporting. Like for similar studies with a different solution approach [46], some other limitations of this methodology should be clarified for future research:

1. Utilizing architectural domain knowledge can prevent acquisition of huge unnecessary data [47]. A pre-scan known from autonomous driving [48] which identifies the object and its bounding box would be helpful. Furthermore, by using a combined approach, the identification and evaluation of labels could reduce time for recognition and avoid some error by misrecognition in case of similarity between two or more objects [49]. Prevention of outliers (undesired reflexion) would reduce the amount of data.
2. Improve the robustness and the usability of data acquisition procedure by utilization of low-cost and mobile devices (e.g. augmented reality glasses, 360-degree camera) [11]. Currently, object recognition based on such clouds provides insufficient results due to both discontinuous cloud density and poor position accuracy (Chap. 4).
3. Reduce or prevent the impact of occlusion by improved segmentation which may consider the semantic relationships between the objects. Most likely, several algorithms for different object types will be necessary.
4. Improve the robustness of recognition methods by better training and continuous learning. Combination of a primary and a secondary recognition algorithm would prevent a misinterpretation of an object, e.g. in case of occlusion [24].
5. Extend the content and usability of the object library by strong involvement of industry circles and continuous improvement. That could be implemented as an augmentation of ModelNet library. Combination with further libraries (CAD standard part libraries) could improve the results and dramatically save the preparation costs.
6. While the training is the most expensive step with manual assistance, further efforts must be spent to reduce the overall effort. This can be conducted by automatic procedures.
7. Better acquisition of non-geometric information. A similar approach for this portion of data is necessary to acquire and collect approximately 400 types of data which occur in a factory and are relevant for planning tasks [50]. This area of reverse engineering is almost not investigated [51].
8. Better integration of singular steps [52]. Basically, it is about reverse engineering, which was included in leading CAD systems a decade ago. Assuming a fast and reliable recognition algorithm, it could be implemented as a module in a CAD system and integrated into PLM as an object recognition module [53].

In order to successfully contend with the diversity of facilities, existing object recognition methods must be combined, scaled and created new ones. In order to facilitate this mission, three limitations of existing frameworks were rectified:

(1) understanding fundamental function,
(2) inferring scope of effectiveness, and
(3) performing an extensive and quantitative comparative performance evaluation.

The two fundamental functions performed by object recognition methods are (1) feature standardization and (2) discriminator distillation. Understanding these fundamental functions enables researchers to accurately interpret parts of methods as being vital or unnecessary [46].

Bottom-up or system-based concepts like this presented here offer better insights into the system to be developed and support the maximum possible (and maximum meaningful) parameterization of the system, they are associated with considerable expenditure of time and thus high development costs, since the interfaces between all virtual and real components must be clearly structured and defined [54]. The possibilities and application of the Digital Twin are additionally limited by the infrastructure and software used, including computing power which significantly exceeds typical CAD applications [18].

The legitimacy of object reconstruction with semantic enrichment needs to be justified in terms of cost–benefit [11, 13]. This research was focused on fast object recognition. If this attempt is successful, the justification is easy. However, the drawbacks must be considered to prevent use case with no recognition (e.g. on significant occlusion). Theoretically, in highly dense industrial spaces data acquisition could be made impossible both by a missing accessibility though the acquisition device and the poor recognition caused by incomplete point cloud and occlusion [33]. Theoretically, such obstacles can only be resolved by an incremental scanning (e.g. by a non-limited mobile device) [55].

Because the Digital Twin is subject to frequent changes, the question arises of how its consecutive updates (e.g. replacement for a machine in a larger hall) can be conducted without restraint to repeat the entire process described above [56]. A simple procedure can be therefore developed in which the user company is enabled to undertake the initial part of the update process itself. For this purpose, an application is being developed with which the partial spaces of the hall to be updated in the Digital Twin can be scanned using the camera of a standard smartphone. The application will guide the user during the recording process, so that handling can be carried out quickly and easily. The scan recorded with the application is further processed automatically to a high degree and the Digital Twin is updated. With the extension of the object recognition for the part of a model to be developed, the existing Digital Twin is analysed and changes are automatically overlayed. This can be, e.g. used to integrate behaviour models of a CNC machine.

8.8 Conclusions and Outlook

To enhance intelligent manufacturing, comprehensive discrete event simulations based on the Digital Twin paradigm represent one of the key technological solutions. In the presence of big or complex data, the Digital Twin plays a crucial role in achieving the convergence between physical and virtual spaces [57]. The intention of this chapter is to advances the realm of generation of a Digital Twin in the highly automated way in a built environment with complicated scenes (e.g. indoor environments with repetitive, irregular-shaped objects, and noisy measurement data as input). With this, the Digital Twin and the use of discrete event simulation provide manufacturing companies improvement potential for production systems leading to various benefits such as improved flexibility and cost savings.

The technological base for generation of Digital Twin is a seamless, robust, (semi-)automatic workflow of primarily standard, modular components with low user assistance. In this chapter, several options for this workflow were discussed as well as the parameters for their selection. It was demonstrated how the overall procedure for the automated generation of a Digital Twin yields repeatable results, which data and information is required, and how it is stored in a useful and process-oriented way. Sensitivity analysis, pitfalls and drawbacks were addressed too with hints for their resolution. Finally, it was proven that a generation process based on scanning and object recognition can run successfully and much faster than by manual rework and fulfil requirements for simulation and process planning.

Another interesting area of research is adoption of these findings in the 3D facility layout problem towards a multi-criteria resolution [5]. Configuration software providers seek to define the fine configuration of real workshops. They essentially consider the spatial layout of the facilities, taking into account the constraints of handling products, tools and equipment but also ergonomics. The use of a calculation method could lead to reduce the size of the solution space from the early stages of the study and, thus, reducing the effort required to finely define the workshop configuration [2].

The Digital Twin is the basis for each subsequent step. In the sales area, the manufacturer can offer batch size one. On the basis of their current capacity utilization situation, them can give the customer an exact delivery date. Furthermore, them can make an expensive offer if the capacity is high, and optimize the price if the capacity is low. With the Digital Twin, them can simulate production layout changes and react more quickly to the market situation through material flow changes. An intuitive visualization of the plant status for a transparent representation of the production helps to further save costs [58].

Likewise, the overall process must be further expanded and obstacles, such as the occlusion of objects, must be removed. If this is successful, the efficient generation of a Digital Twin can be provided as a service to companies. With the procedure described, which covers all required use cases, it is possible that a Digital Twin in the built environment is created in few days and with small effort. Likewise, an update of existing models is feasible. The companies do not necessarily need their own experts with programming skills, but are supplied with a complete simulation model through the comprehensive service concept [58]. Correspondingly, planning can be carried out more efficiently and flexibly, and productivity and quality increases are created.

References

1. Eriksson A, Sedelius E, Berglund J, Johansson B (2018) Virtual factory layouts from 3D laser scanning—a novel framework to define solid model requirements. Procedia CIRP 76:36–41. https://doi.org/10.1016/j.procir.2018.01.013
2. Besbes M, Zolghadri M, Costa Affonso R, Masmoudi F, Haddar M (2020) 3D facility layout problem. J Intell Manuf. https://doi.org/10.1007/s10845-020-01603-z
3. Lindskog E, Berglund J, Vallhagen J, Johansson B (2016) Layout planning and geometry analysis using 3D laser scanning in production system redesign. Procedia CIRP 44:126–131. https://doi.org/10.1016/j.procir.2016.02.118
4. Tolio T, Sacco M, Terkaj W, Urgoa M (2013) Virtual factory: an integrated framework for manufacturing systems design and analysis. Procedia CIRP 7:25–30. https://doi.org/10.1016/j.procir.2013.05.005
5. Iqbal M, Hashmi MSJ (2001) Design and analysis of a virtual factory layout. J Mater Process Technol 118(1–3):403–410. https://doi.org/10.1016/S0924-0136(01)00908-6
6. Kuhn O, Liese H, Stjepandić (2011) Methodology for knowledge-based engineering template update. In: Cavallucci D, Guio R, Cascini G (eds) Building innovation pipelines through computer-aided innovation. Springer, Berlin Heidelberg, pp 178–191. https://doi.org/10.1007/978-3-642-22182-8_14
7. Nåfors D, Lindskog E, Berglund J, Gong L, Johansson B, Vallhagen J (2017) Realistic virtual models for factory layout planning. In: IEEE 2017 winter simulation conference (WSC). https://doi.org/10.1109/WSC.2017.8248107
8. Li CT, Cheng JCP, Chen K (2020) Top 10 technologies for indoor positioning on construction sites. Autom Constr 118:103309. https://doi.org/10.1016/j.autcon.2020.103309.
9. Geng Z, Bidanda B (2017) Review of reverse engineering systems—current state of the art. Virtual Phys Prototyp 12(2):161–172. https://doi.org/10.1080/17452759.2017.1302787
10. Šagi G, Lulić Z, Mahalec I (2015) Reverse engineering. In: Stjepandić J et al (eds) Concurrent engineering in the 21st century: foundations, developments and challenges. Springer International Switzerland, pp 319–353. https://doi.org/10.1007/978-3-319-13776-6_12
11. Wang K, Kim M-K, Applications of 3D point cloud data in the construction industry: a fifteen-year review from (2004) to 2018. Adv Eng Inform 39(2019):306–319. https://doi.org/10.1016/j.aei.2019.02.007
12. Lu Q, Chen L, Li S, Pitta M (2020) Semi-automatic geometric digital twinning for existing buildings based on images and CAD drawings. Autom Constr 115:103183. https://doi.org/10.1016/j.autcon.2020.103183.
13. Ma Z, Liu S (2018) A review of 3D reconstruction techniques in civil engineering and their applications. Adv Eng Inform 37:163–174. https://doi.org/10.1016/j.aei.2018.05.005
14. Bondar S, Salem B, Stjepandić J (2018) Indoor Object reconstruction based on acquisition by low-cost devices. Adv Transdiscipl Eng 7(2018):113–122. https://doi.org/10.3233/978-1-61499-898-3-113
15. Pučko Z, Šuman N, Rebolj D (2018) Automated continuous construction progress monitoring using multiple workplace real time 3D scans. Adv Eng Inform 38:27–40. https://doi.org/10.1016/j.aei.2018.06.001
16. Fathi H, Dai F, Lourakis M (2015) Automated as-built 3D reconstruction of civil infrastructure using computer vision: achievements, opportunities, and challenges. Adv Eng Inform 29(2):149–161. https://doi.org/10.1016/j.aei.2015.01.012
17. Ríos J, Mas Morate F, Oliva M, Hernández JC (2016) Framework to support the aircraft digital counterpart concept with an industrial design view. Int J Agile Syst Manage 9(3):212–231. https://doi.org/10.1504/IJASM.2016.079934.
18. Adamenko D, Kunnen S, Pluhnau R, Loibl A, Nagarajah A (2020) Review and comparison of the methods of designing the Digital Twin. Procedia CIRP 91:27–32. https://doi.org/10.1016/j.procir.2020.02.146

19. Fukuda S, Lulić Z, Stjepandić J (2013) FDMU-functional spatial experience beyond DMU? In: Proceedings of the 20th ISPE international conference on concurrent engineering, CE 2013, IOS Press, Amsterdam, pp 431–440. https://doi.org/10.3233/978-1-61499-302-5-431
20. Biahmou A, Emmer C, Pfouga A, Stjepandić J (2016) Digital master as an enabler for industry 4.0. Adv Transdiscipl Eng 4:672–681. https://doi.org/10.3233/978-1-61499-703-0-672
21. Pfouga A, Stjepandić J (2018) Leveraging 3D geometric knowledge in the product lifecycle based on industrial standards. J Comput Des Eng 5(1):54–67. https://doi.org/10.1016/j.jcde.2017.11.002
22. Bondar S, Ruppert C, Stjepandić J (2014) Ensuring data quality beyond change management in virtual enterprise. Int J Agile Syst Manage 7(3/4):304–323. https://doi.org/10.1504/IJASM.2014.065346
23. Princeton Modelnet Repository. https://modelnet.cs.princeton.edu/#. Accessed 30 Nov 2020
24. Salem B, Stjepandić J, Stobrawa S (2019) Assessment of methods for industrial indoor object recognition. Adv Transdiscipl Eng 10:390–399. https://doi.org/10.3233/ATDE190145
25. Emmer C, Fröhlich A, Stjepandic J (2013) Advanced engineering visualization with standardized 3D formats. In: IFIP advances in information and communication technology, vol 409. Springer, Berlin Heidelberg, pp 584–595. https://doi.org/10.1007/978-3-642-41501-2_58
26. www.opendesc.com. Accessed 16 November 2020
27. Ostrosi E, Stjepandić J, Fukuda S, Kurth M (2014) Modularity: New trends for product platform strategy support in concurrent engineering. Adv Transdiscipl Eng 1:414–423. https://doi.org/10.3233/978-1-61499-440-4-414
28. Sommer M, Stjepandić J, Stobrawa S, von Soden M (2021) Automated generation of a digital twin in manufacturing for a built environment using scan and object detection. J Ind Inf Integr (in press)
29. Ochmann S, Vock R, Klein R (2019) Automatic reconstruction of fully volumetric 3D building models from oriented point clouds. ISPRS J Photogramm Remote Sens 151(2019):251–262. https://doi.org/10.1016/j.isprsjprs.2019.03.017
30. Rusu RB, Cousins S (2011) 3D is here: Point Cloud Library (PCL). In: 2011 IEEE international conference on robotics and automation, ICRA 2011, Article number 5980567. https://doi.org/10.1109/ICRA.2011.5980567
31. Lehtola VV, Kaartinen H, Nüchter A, Kaijaluoto R, Kukko A, Litkey P, Honkavaara E, Rosnell T, Vaaja MT, Virtanen J-P, Kurkela M, El Issaoui A, Zhu L, Jaakkola A, Hyyppä J (2017) Comparison of the selected state-of-the-art 3D indoor scanning and point cloud generation methods. Remote Sens 9:796. https://doi.org/10.3390/rs9080796
32. Moon S, Chung S, Kwon J, Seo J (2019) Shin, comparison and utilization of point cloud generated from photogrammetry and laser scanning: 3D world model for smart heavy equipment planning. Autom Constr 98:322–331. https://doi.org/10.1016/j.autcon.2018.07.020
33. Garcia-Garcia A, Garcia-Rodriguez J, Orts-Escolano S, Oprea S, Gomez-Donoso F, Cazorla M (2017) A study of the effect of noise and occlusion on the accuracy of convolutional neural networks applied to 3D object recognition. Comput Vis Image Underst 164:124–134. https://doi.org/10.1016/j.cviu.2017.06.006
34. Melcher D, Küster B, Stonis M, Overmeyer L (2019) Fabrik-und Produktionsplanung im digitalen Modell durch automatisierte Punktwolkenverarbeitung. In: Logist J Proceed 2019. https://doi.org/10.2195/lj_Proc_melcher_de_201912_01
35. Rebolj D, Pučko Z, Čuš Babič N, Bizjak M, Mongus D (2017) Point cloud quality requirements for Scan-vs-BIM based automated construction progress monitoring. Autom Constr 84:323–334. https://doi.org/10.1016/j.autcon.2017.09.021
36. Bicocchi N, Cabri G, Mandreoli F, Mecella M (2019) Dynamic digital factories for agile supply chains: an architectural approach. J Ind Inf Integr 15:111–121. https://doi.org/10.1016/j.jii.2019.02.001
37. Bondar S, Shammaa A, Stjepandić J, Tashiro K (2015) Advances in parameterized CAD feature translation. Adv Transdiscipl Eng 2:615–624. https://doi.org/10.3233/978-1-61499-544-9-615
38. Yang W, Fu C, Yan X, Chen Z (2020) A knowledge-based system for quality analysis in model-based design. J Intell Manuf 24. https://doi.org/10.1007/s10845-020-01535-8

39. Gürdür D, El-khoury J, Nyberg M (2019) Methodology for linked enterprise data quality assessment through information visualizations. J Ind Inf Integr 15 September 2019, pp 191–200. https://doi.org/10.1016/j.jii.2018.11.002
40. Sun J, Hiekata K, Yamato H, Nakagaki N, Sugawara A (2014) Virtualization and automation of curved shell plates manufacturing plan design process for knowledge elicitation. Int J Agile Syst Manage 7(3/4):282–303. https://doi.org/10.1504/IJASM.2014.065348
41. Schuh G, Rebentisch E, Riesener M, Ipers T, Tönnes C, Jank MH (2019) Data quality program management for digital shadows of products. Procedia CIRP 86:43–48. https://doi.org/10.1016/j.procir.2020.01.027
42. Kawashima K, Kanai S, H. (2014) As-built modeling of piping system from terrestrial laser-scanned point clouds using normal-based region growing. J Comput Des Eng 1(2013):13–26. https://doi.org/10.7315/JCDE.2014.002
43. Agapaki E, Brilakis I (2020) CLOI-NET: class segmentation of industrial facilities' point cloud datasets. Adv Eng Inf 45:101121. https://doi.org/10.1016/j.aei.2020.101121
44. Inyoung Oh, Hee Ko K (2020) Automated recognition of 3D pipelines from point clouds. Vis Comput. https://doi.org/10.1007/s00371-020-01872-y
45. Quintana B, Prieto SA, Adán A, Vázquez AS (2016) Semantic scan planning for indoor structural elements of buildings. Adv Eng Inform 30:643–659. https://doi.org/10.1016/j.aei.2016.08.003
46. Czerniawski T, Leite F (2020) Automated digital modeling of existing buildings: a review of visual object recognition methods. Autom Constr 113. https://doi.org/10.1016/j.autcon.2020.103131
47. Dekhtiar J, Durupt A, Bricogne M, Eynard B, Rowson H, Kiritsis D (2018) Deep learning for big data applications in CAD and PLM—research review, opportunities and case study. Comput Ind 100:227–243. https://doi.org/10.1016/j.compind.2018.04.005
48. Liu D, Cui Y, Chen Y, Zhang J, Fan B (2020) Video object detection for autonomous driving: motion-aid feature calibration. Neurocomputing 409:1–11. https://doi.org/10.1016/j.neucom.2020.05.027
49. Otero R, Lagüel S, Garrido I, Arias P (2020) Mobile indoor mapping technologies: a review. Autom Constr 120:103399. https://doi.org/10.1016/j.autcon.2020.103399
50. da Silva Serapião Leal G, Guédria W, Panetto H (2019) An ontology for interoperability assessment: a systemic approach. J Ind Inf Integr 16:100100. https://doi.org/10.1016/j.jii.2019.07.001
51. El-khoury J, Berezovskyi A, Nyberg A (2019) An industrial evaluation of data access techniques for the interoperability of engineering software tools. J Ind Inf Integr 15(2019):58–68. https://doi.org/10.1016/j.jii.2019.04.004
52. Wagner L (2020) Kein Digital Twin ohne digitale Durchgängigkeit. Zeitschrift für den wirtschaftlichen Fabrikbetrieb, Band 115, April 2020, pp 43–45. https://doi.org/10.3139/104.112315
53. Wognum N, Bil C, Elgh F, Peruzzini M, Stjepandić J, Verhagen WJC (2019) Transdisciplinary systems engineering: implications, challenges and research agenda. Int J Agile Syst Manag 12(1):58–89. https://doi.org/10.1504/IJASM.2019.098728
54. Ruppert T, Abonyi J (2020) Integration of real-time locating systems into digital twins. J Ind Inf Integr 20. https://doi.org/10.1016/j.jii.2020.100174
55. Xue F, Lu W, Chen K, Webster CJ (2019) BIM reconstruction from 3D point clouds: a semantic registration approach based on multimodal optimization and architectural design knowledge. Adv Eng Inf 42. Article number 100965. https://doi.org/10.1016/j.aei.2019.100965
56. Ladj A, Wang Z, Meski O, Belkadi F, Ritou M, Da Cunha C (2020) A knowledge-based Digital Shadow for machining industry in a Digital Twin perspective. J Manuf Syst. https://doi.org/10.1016/j.jmsy.2020.07.018
57. Shen H, Bednarz T, Nguyen H, Feng F, Wyel T, Hoek PJ, Lo EHS (2019) Information visualisation methods and techniques: state-of-the-art and future directions. J Ind Inf Integr 16:100102. https://doi.org/10.1016/j.jii.2019.07.003

58. Bondar S, Potjewijd L, Stjepandić J (2013) Globalized OEM and tier-1 processes at SKF. In: Stjepandic et al (eds) Concurrent engineering approaches for sustainable product development in a multi-disciplinary environment. Springer, London, pp789–800. https://doi.org/10.1007/978-1-4471-4426-7_67

Chapter 9
Design of Simulation Models

Sebastian Stobrawa, Gina Vibora Münch, Berend Denkena, and Marc-André Dittrich

9.1 Introduction

While the previous chapters have described the basics of a framework for a Digital Twin and how the required objects and parameters can be acquired, this chapter focuses on the setup of a virtual model. The virtual model is an integral part for the Digital Twin, since it operates on this model, and provides the digital representation of the real system. The setup of a model is usually complex as well as time-consuming and requires manual work by a suitably qualified person. According Fritzsche et al. [1], setting up a new simulation model takes approximately 54 days. The example is typical for simulation studies and shows the high effort and high costs. However, it must be considered that the effort for the creation depends considerably on the degree of complexity. Nevertheless, an indication of the effort required is given here. Therefore, the automation of the setup process offers great cost savings, which could lead to a wider diffusion of the Digital Twin technology.

For a (semi-)automated generation process of virtual models of production systems there are further use cases. For example, Stobrawa et al. [2] explains how an automatic model generation is used for the validation of control approaches. In Addition, Kikolski [3] presents an approach to generate different scenarios from a

S. Stobrawa (✉) · G. V. Münch · B. Denkena · M.-A. Dittrich
Institut Für Fertigungstechnik Und Werkzeugmaschinen, Leibniz Universität Hannover, Universität 2, 30823 Garbsen, Germany
e-mail: stobrawa@ifw.uni-hannover.de

G. V. Münch
e-mail: vibora-muench@ifw.uni-hannover.de

B. Denkena
e-mail: denkena@ifw.uni-hannover.de

M.-A. Dittrich
e-mail: dittrich@ifw.uni-hannover.de

J. Stjepandić et al. (eds.), *DigiTwin: An Approach for Production Process Optimization in a Built Environment*, Springer Series in Advanced Manufacturing,
https://doi.org/10.1007/978-3-030-77539-1_9

simulation model. Since every use of a digital twin or a simulation model starts with the setup, all these processes benefit from a simplification of the setup process.

This section therefore intends to simplify the generation process. This will be achieved by introducing a generic architecture for the virtual representation of production systems. This architecture provides a description of the structure and can be transferred to specific applications in any simulation software. Since the present book shows a concrete implementation, i.e. an exemplary and prototypical application, a limitation of the use case and specification of the basic conditions will follow. Subsequently, the procedure for (semi-)automated setup is presented in this special case. Finally, specific results will be given, especially with illustrations from the simulation environment.

9.2 Ontology of Production Systems

The model structure for the virtual representation of a production system requires a formal description of the system, including the elements and their interdependencies. This generic approach is essential for the generation of a Digital Twin, as described in this book, in order to enable the transfer of the setup process of the digital twin for any production system. In this context, an ontology is suitable as formal description. In the following, an example will show how such an ontology can be set up.

9.2.1 Ontology-Based Information System for Generating Simulations of Production Systems

Simulation environments apply standardized modules which are adapted to the individual production system by means of production data. For this purpose, production systems are described as a system consisting of different system components and their interactions with each other [4]. In this definition the system components represent objects. To these objects each real instance of a production system can be assigned. The standardised modules are adapted to the individual characteristics of a company-specific production system by means of attributes. For setting certain values for these attributes, information from the respective production system are required. In order to structure the information, and to provide it in a suitable data exchange format, different forms of information technologies are applied. A standard used for this form of information structuring is the Core Manufacturing Simulation Data (CMSD) standard. By setting up superordinate information classes, production data are processed in such a way that they can be assigned to the standardised modules that represent a specific class of objects in the simulation of a production system.

9.2.2 *Information System for the Description of Production Systems*

To map a production system in a simulation environment properties of the system and the system components are derived from production data. The VDI—The Association of German Engineers defines three types of production data that are relevant for simulations of production systems [5]. Figure 9.1 shows relevant data for the simulation. The VDI defines system load data, organizational data and technical data as input for the simulation.

Technical data includes data of the factory layout, production resource characteristics and production system topology. Organizational data includes data about the structuring of processes and operations as well as planning and scheduling strategy, for example the resource allocation. System load data includes data regarding orders, batch sizes or scheduling decisions [5]. In addition to the pure definition of the necessary data, the preparation of data is to determine for the automated generation of simulation models. In order to take the user (programmer) out of the simulation generation process, the manual derivation of information from production data and the parameterisation of the simulation model, based on this, must be specified in a machine-processable form. Otherwise, no representation of the properties and behaviour of the production system is possible without human interpretation. This means that the information system must enable the assignment of the components that actually exist in production to the pre-programmed simulation objects and the assignment of the real attribute characteristics to the attributes of the simulation objects. Ontologies offer the possibility to build such an information system. For a better understanding, a brief summary about information systems and ontologies is given in the following.

Various information technologies are applied to set up information systems. The term information technology refers to technical systems that are deployed to acquire, store and process information [6]. Also, communication is a task of information technologies. A distinction is made between different information technologies in terms of the content they can represent [6]. Simple information technologies, such as

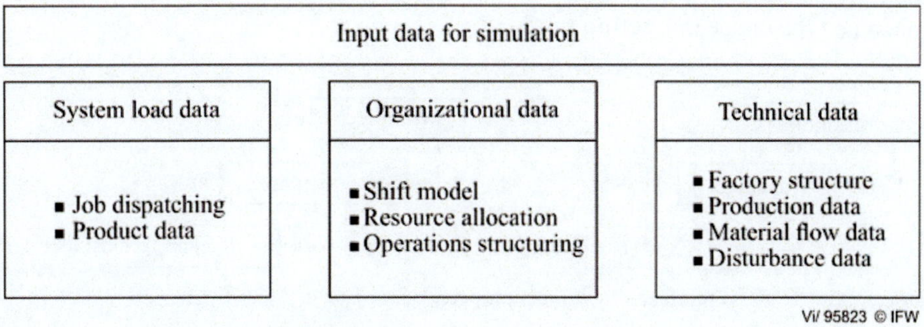

Fig. 9.1 Input data for simulation models described by VDI (in reference [5])

lists, enable data to be structured. More complex structures, such as tree or network structures, are provided to associate data with each other. For the automatic generation of simulation models, ontologies are used by these complex information systems. They can represent a high level of information content by store data:

- arrange data hierarchically,
- represent similarity relations,
- represent synonym relations,
- depict relationships,
- specify areas of validity and connect data to information by logical links [6].

Ontologies are used to structure a set of terms and relate them to each other. This makes them suitable for representing the relationships of the production components with regard to the production process. Likewise, for mapping the components to higher-level standardised object modules that are utilized in the simulation. The terms are divided into so-called classes, as in the example shown in Fig. 9.2. These classes, for example "workstation", have different attributes that make them identifiable [7]. Exemplary attributes are the ID (name) in form of a string or a current status ("working", "disturbed" or "paused"). A term or a class can have subclasses. These subclasses have the attributes of the superordinate class and additional attributes which distinguish them from the superordinate class. Instances can be assigned to the classes [7]. These instances have individual expressions of the attributes of a class. They represent real objects, like the turning machine CTX520L in the example. Relationships between classes represents so-called object properties. An example for this object properties is shown Fig. 9.2. It exists a functional relationship between the class "workpiece feature" and the class "machines". This relationship represent the information that a workpiece feature is produced by a machine.

Only the relationship is represented in the ontological information. The function of the relationship must be interpreted individually [7]. So, the expression "is produced by" must be interpreted. The example shows that the ontology only specifies which object is part of a function by linking terms. In the example the machine is used to create a workpiece feature. Data properties are another aspect of ontological information system. The properties represent defined attributes of the instances in the form of values or expressions. Examples of such specific attributes are a specific machine type or specific set-up times.

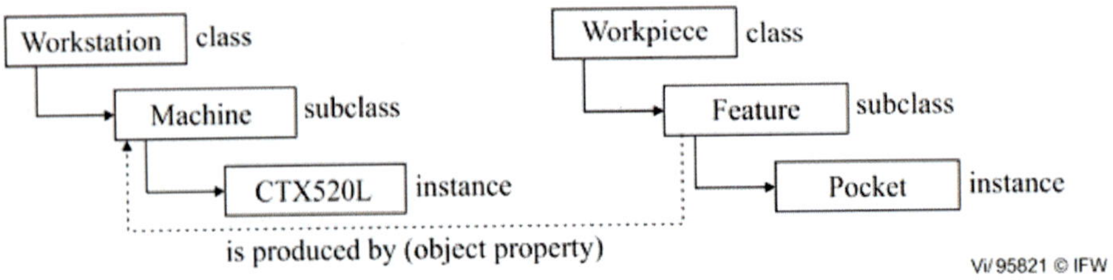

Fig. 9.2 Example of the basic hierarchy of classes and instances

The basic structure of ontologies, which stores information in classes associated with attributes and instances, becomes machine-processable through the semantics applied. The semantics specify in a machine-processable way how the data in the ontology is accessed and how the information resulting from the structuring of the data is to be interpreted [5].

Ontologies are already deployed to structure information for the automatic generation of simulation models. One standard that encompasses an object-oriented information system is the CMSD. The CMSD standard merges the relevant production data defined by the VDI in a superordinate way. The function-oriented structuring of the VDI is translated in an object-oriented way [5]. So, the information is assigned to the objects: part (workpiece), resource (workstation, equipment) and production operation [8]. In addition, the standard defines additional classes in which supporting information for the creation of simulation models is specified, see Fig. 9.3. The classes represents the components of the real manufacturing system, on general level and on instance level.

Part information describe the workpiece. Attributes of the workpiece are the current processing status, the current booking point in the simulations of a production system or the name (workpiece ID) for example. The class record information from all three information areas of the VDI definition. Resource information describe workstations, employees and other equipment of the production system. Attributes of the resources are the current state or the individual name (ID) for example. Especially in this class, the technical data are represented. Process data are applied in the production planning and production operation class to describe the manufacturing process. This includes data about the specific manufacturing steps and organisational boundary conditions. In addition, this information class defines different jobs with the organisational and workload data from the VDI model. The Layout class represent information for the resource location and standardise resource symbols for the simulation. With the superordinate class "support" the data format, the data structure and data exchange are provided by the CMSD Standard. The standard is based on the Unified Modelling Language (UML) and has also been implemented with the

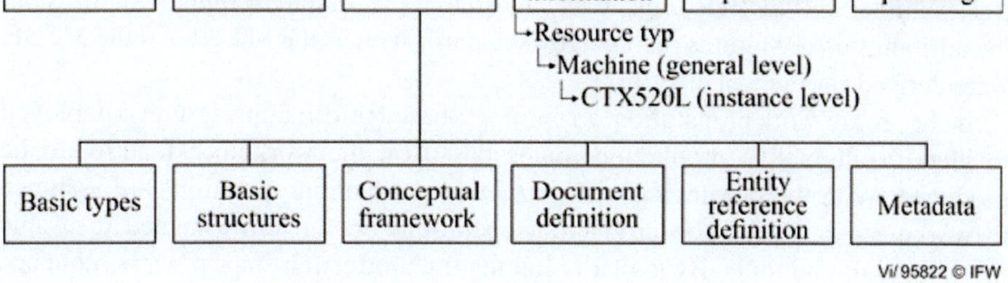

Fig. 9.3 Classes of the CMSD information model [8]

RDF/XML syntax [7]. The Extensible Mark-up Language (XML) is a text-based data format which can be read and edited by computers. XML enables the structuring of documents [9]. For this purpose, the mark-up language works with so-called tags, whose appearance is standardised. The meaning of the tag is not predefined. Tags can be harnessed to describe, store and exchange data. Hierarchies can be mapped with tags by nesting [9]. So, the XML is deployed to structure the production system into classes and subclasses. The Resource Description Framework Schema (RDF) is used to map properties and relationships of resources in a machine-readable form. The fact that the standard is based on the XML/RDF syntax simplifies the exchange of data between the real production system and the simulation environment [5]. Production data are usually collected in RDF/XML based syntax in the shop floor. The data is often stored and made available to the information system using an SQL database [10–12].

There are two different ways to use structured data for the generation of simulation models. Internal approaches follow the implementation in the simulator (the software in which the simulation model is built). External approaches create an algorithm or a script in another programming environment. In this environment, the algorithm generates a source code for the simulation [5]. The algorithm is part of the next chapters of the book.

9.2.3 Exemplary Applications

The CMSD standard was developed to provide production data in a uniform data structure. This unified data structure is applied by various approaches to create a material flow simulation. The approaches differ in whether they prepare the data for a programmer or use it to automatically create material flow simulations. FOURNIER [13] exploits the standard to build a translator for the programmer that provides the data from the CMSD in a simplified way. He applies the XML syntax to read the data from the CMSD and make it available for programming in the *FlexSim* material flow software. BERGMANN et al. [5] use the standard for the automated generation of a simulation model. They apply the standard to process the data automatically and to build an executable simulation model of the production in the material flow simulation software *Plant Simulation* [5]. In the project, the enterprise resource planning system (for example SAP ERP) is utilized as data source [5]. In the exemplary application, the attributes of key figures recorded by standard, such as the MTTR and the MTBF, were derived for the instances [5].

In the area of work preparation, an ontology-based information system is deployed to identify alternative production routes based on the workpiece features to be produced. With the ontological class structure production operations are assigned to workpiece features. The production operations are in turn assigned to usable machine tools and tools. By logically linking the production steps process sequence alternatives are identified and displayed as non-linear work plan [12]. Further developments utilize this information to check the derived work plans with digital twins

[14]. These routes are already being evaluated according to various criteria [12, 15]. For example, a cost estimate is used to make decisions for production routes. In all approaches the information system provides the information. The actual calculation and thus the determination of results is done by calculation with standardised equations, for example at process time determination or by mapping in a material flow simulation environment.

In publications the existing approaches for the automatic generation of material flow simulations are compared and research gaps are highlighted (see for example BERGMANN et al. [16]). This shows that the lack of standardisation is still a major problem. That makes it difficult to compare different research results with each other. Here a standard like CMSD can close the gaps. Like a kind of gateway, the information system can adapt data from production to the logic of the simulation. Or combine the logic of different partial solutions for the generation of simulation models. Due to its class structure, the ontology-based information system offers the possibility of transferring various functions, such as a scheduling procedure.

Currently existing methods for generating simulation models are semi-automated methods. A planner or programmer is still required to manually model certain features of the behaviour and characteristics of the production system [16]. This greatly increases the effort required to generate simulation models. That leads to the fact that, despite the manual extension, only partial aspects of a production system are usually modelled. This means that only a part of the possibilities of existing simulation environments can be deployed. As shown in Fig. 9.4, simulation models can be applied to represent various components of a real production system.

In order to harness all the functions of material flow simulation, a comprehensive database is necessary. At the same time, the effort involved in generating the simulation model must be kept as low as possible. This included, that the data is transferred to an information system with as low effort as possible. Existing information systems of the production system, for example a Manufacturing Execution System (MES), enable this in particular for the generation of resources, orders or products and for production planning and control in a simulation environment. Layout information

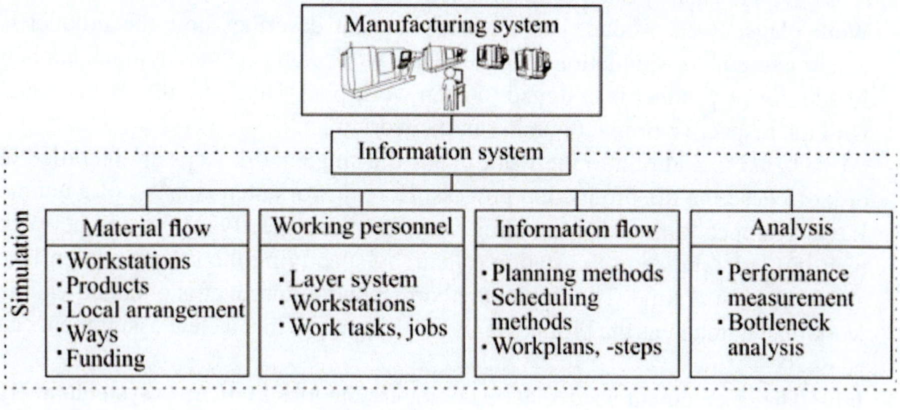

Fig. 9.4 Transfer to the components of a simulation

is usually not yet provided by the systems. The project addresses this gap. In the following, it is shown how information on the layout is provided by means of a scan of the production and be used to generate a simulation model. An automated process greatly reduces the effort required to create the simulation.

9.3 Data Model

In this section, the described ontology is now used to derive an architecture for the model structure. This architecture corresponds to the structure of the elements in the model and also to the data model. For this reason, the architecture is represented in the form of an entity relationship model (ERM). The advantage of this approach for description is that the ERM can be transferred directly into a relational database, which is utilized for the program flow for the model. Thus, the description determines the later procedure of the model and it ensures that the method can be transferred to any production system.

For the purpose of illustration and comprehensibility, a basic structure is first presented. This structure can then be extended in the sense of the ontology described above. For this purpose, the ontology from Fig. 9.4 was reduced to essential elements. The basic structure consists of [17]:

- Stations: This refers to all objects where value is added, i.e. machines, workplaces, plants, etc. The stations must receive a unique identification number (ID) in addition to a label with which the station is named in the production system. The ID is required for the program-technical sequence of the simulation. In addition, the spatial position of the stations is used.
- Products: The products that are produced in a production system are registered at this point. The products also require a unique ID for use in the simulation. In addition, a label is defined. If several products are processed similarly in the simulation, products can be bundled to product families. In this case, each product is assigned to exactly one product family.
- Work plans: Each product has a work plan that describes how the product is manufactured. For simulation purposes, each work plan is given a unique number. In addition, a product is assigned to each work plan. Here, the product is noted with the respective product number in the work plan.
- Work steps: In addition to the work plans, the single work steps are recorded in order to describe the production process. Here, a work step consists of a unique ID, a reference to the corresponding work plan (by specifying the unique work plan ID), a reference to the station (again, by specifying the unique station ID), and a processing time. The link to the corresponding product is obtained via the work plan number, as the product ID is stored here, so that a clear assignment can be made.
- Jobs: The last element needed at this point are the jobs in order to systematically describe a production process. The jobs consist of a unique number, an assigned

product (noted by the product ID), the start time of the job, and the job quantity, i.e. the batch size.

With these elements a basic functionality of the production system can already be enabled. Furthermore, this basic functionality can be extended by the elements described in Fig. 9.4 in the sense of the ontology. Accordingly, this simplified model forms a basis for a simulation that can be extended at any desired time and according to specific requirements. Figure 9.5 shows an ERM that puts this basic structure in context and can be transferred to a relational database.

As mentioned, the basic functionality can be extended as required so that the model correctly represents the real system. For the extension, the ontology was introduced in the first sections of this chapter. At this point, it is not intended to expand on the basic functionality presented, as each extension must be appropriately matched to the real system. In the further course of the chapter, however, it will be described how this basic functionality can be set up automatically and at the end of the section an outlook on reasonable and feasible extensions will be given. It should be emphasised that an extension of the basic functionality usually requires manual adaptation. However, some extensions are already pre-implemented in the solution presented here, so that these can be applied.

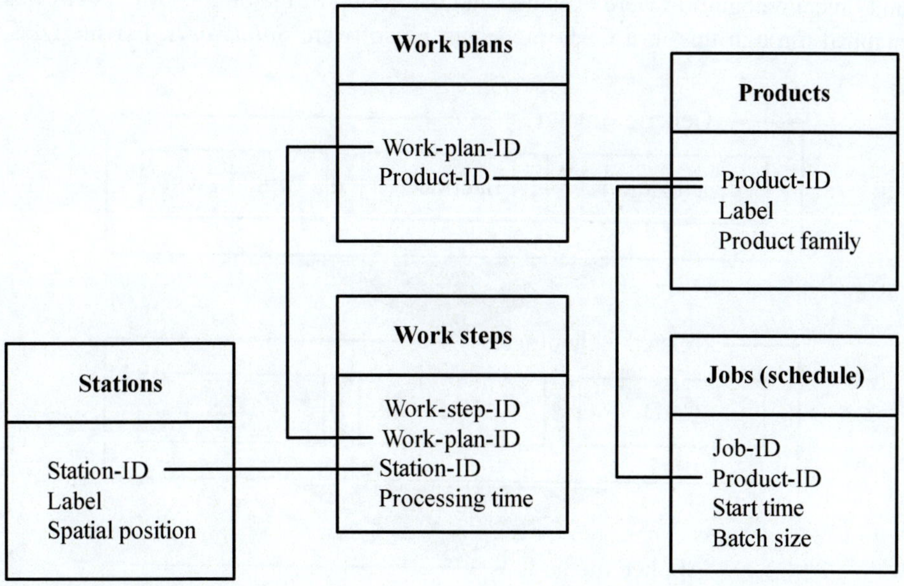

Fig. 9.5 ERM of the data model [17]

9.4 Model Design

Figure 9.6 provides an overview of the basic task. In the upper part of the figure, the task is initially abstracted, which means that the task is presented here in a generic structure. The aim is to set up a Digital Twin from the visualization of a production system, which is recorded by scan (as described in Chap. 4). This process contains of an interface between the visualization and the Digital Twin, which is used to automatically set up the model. More precisely, the design in this approach is represented in a CAD model, generated by the object recognition process (explained in Chap. 8). CAD is a technology for technical visualization and is applied in this context to display the result of the scan and object recognition.

The Digital Twin is targeted in the first step by a discrete event simulation (DES), i.e. a simulation model in the sense of the ontology described in Sect. 9.2. The transfer of the visualization (CAD) into the DES requires a format that distinguishes between objects, that means that it is object-oriented, and is open-source, so that an adaptation to the concrete requirements can be carried out. This format is provided by the XML schema.

A further clarification is given when the corresponding methods for the technologies mentioned are outlined. In the implementation presented in this book, the scan and object recognition were executed and the design of the production system was mapped through this in a CAD model in the software *SolidWorks*. For the DES,

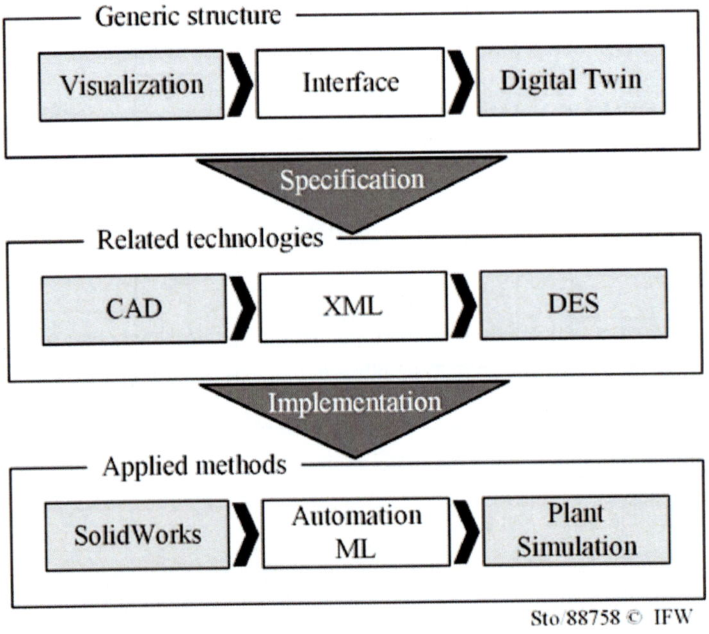

Fig. 9.6 Specification of the subject matter [18]

the software *Plant Simulation* was utilized, which is a standard software for simulation of production systems and enables the development of suitable methods by an internal coding language. For data transfer between SolidWorks and Plant Simulation, Computer-Aided Engineering Exchange (CAEX) was adopted, which is an XML-based format and provides corresponding specifications for the industrial environment. However, this last specification has only been described for the specific implementation presented here. In general, other XML formats (e.g. *Automation ML*) can be transferred from different CAD software (e.g. *CATIA*) to alternative simulation software (e.g. *AnyLogic*).

After this general description, the concrete implementation will be explained. Here, a two-step procedure is applied for the model setup. In the first step, the simulation-capable objects are inserted into the model. The second step provides the graphical representation of the model. This procedure, i.e. the distinction between function and graphical representation, is necessary in order to utilize the pre-defined simulation modules of the simulation software. Accordingly, a functional model is generated that maps the characteristics and behaviour of the production system. The model consists of standard components. However, the model does not correspond to the real system, because the standard components are obviously not equal to the real objects. Therefore, in the second step the functional model is hidden and a graphical depiction of the real system is placed in the model. Since the proportions of both models are equal, the result is a detailed model that contains all simulation-based functionalities. The two steps are explained in the following, successively.

9.4.1 Design of the Functional Model

The functional model includes the information from the data model presented in Sect. 9.3. The resulting model is fully functional, which means that it can be used to perform simulations of the real system. However, it may have to be extended in the specific application if, for example, buffers, transport facilities or similar are required in order to correctly represent the application.

To set up the functional model, the following steps are executed, which can be automated to a large extent:

1. Exporting the object data from the object recognition to an XML-file,
2. Importing the XML file into Plant Simulation,
3. Entering the product data, and
4. Entering the production schedule.

The visualization of the production system as mentioned in the last chapters of this book is generated by the scan and subsequent object recognition and provided as a CAD model. This means that the model is based on a point cloud and the object recognition, that identifies objects of the production system located in this point cloud. Here, a tree structure is used to arrange and correlate different elements of the system. The upstream process was therefore designed in such a way that the tree structure separate objects that must also be distinguished for the simulation model.

The export of the model operates as follows: The tree structure of the model is looped through by the method. The structure is then transferred to an XML file. Since the XML structure can adopt the tree structure of the model, the method is simple and mainly adopts the data that is stored in the model. The process is visualized in Fig. 9.7.

The export method was programmed in a macro for *SolidWorks*. It is open source and can be downloaded from [19]. The macro is written in Visual Basic for Applications (VBA) programming language and must be triggered manually in SolidWorks while a model is open. An XML file is the direct result of the macro and must be saved locally. In general, the XML format uses descriptors that are marked by the symbols <>, just as in the HTML program language. Hence the general structure is:

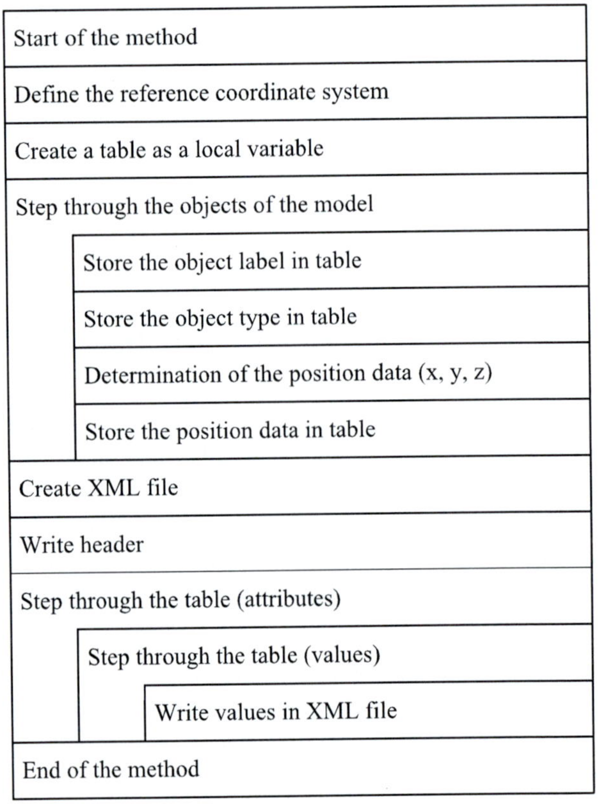

Fig. 9.7 Representation of the process in a Nassi-Shneiderman diagram [18]

```
<?xml version="1.0" encoding="UTF-8" standalone="yes"?>
<eMPlantTable xmlns:pt="urn:schemas-tecnomatix-com:eM-Plant">
    <Table pt:Name="X" pt:XDim="4" pt:YDim="-1">
            <ColumnIndex pt:Type="string">
                    <Cell>Type</Cell>
                    <Cell>X</Cell>
                    <Cell>Y</Cell>
            </ColumnIndex>
            <Column pt:Type="string">
                    <Cell>machine</Cell>
                    <Cell>machine</Cell>
                    <Cell>conveyor</Cell>
            </Column>
            <Column pt:Type="integer">
                    <Cell>10</Cell>
                    <Cell>10</Cell>
                    <Cell>30</Cell>
            </Column>
            <Column pt:Type="integer">
                    <Cell>20</Cell>
                    <Cell>40</Cell>
                    <Cell>20</Cell>
            </Column>
    </Table>
</eMPlantTable>
```

For the introduced application, the <item> stands for an object in the model. Attributes are assigned to each object. These attributes may differ for various objects, depending on the properties of those objects. For example, the processing time can be stored for machines, while for a conveyor, the length is stored in the XML file. However, it is also feasible to store the same attributes for both of the above-mentioned example objects. Here, for instance, the locations in the model are transmitted in X and Y coordinates. Analogous to this procedure, the structure of the XML interface was defined for each object. This procedure enables the transfer of data that is needed again for the basic structure of the simulation model.

Assuming that the model creation is not to be carried out by a specialist, a user interface for importing the XML file was built in the simulation software. This enables a decoupling point to be set to the customer for a later service after object recognition. The user interface is shown in Fig. 9.8. Here, the path to the XML file is simply to be entered. Then, by activating a button, a method for import is started and the model is built automatically.

The import of the XML file into the simulation software is set up in a way that all values from the XML file are first written into an internal table of the simulation software. This procedure is reasonable because it decreases the computing time. The simulation software accesses the XML file only once and extracts all data, which

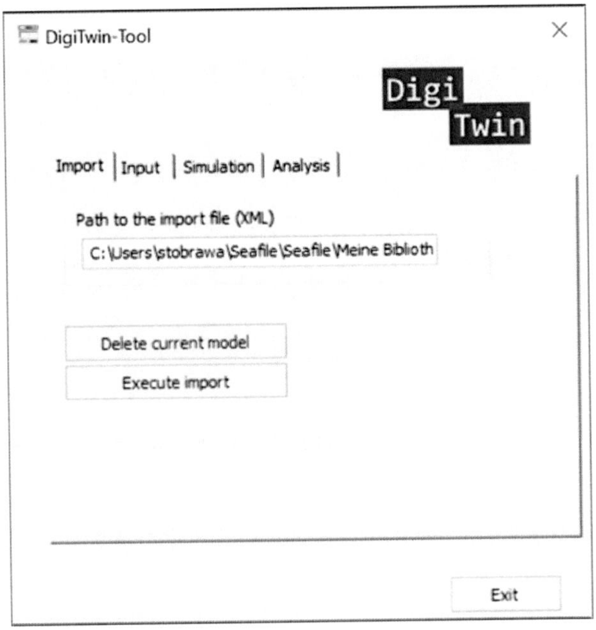

Fig. 9.8 User interface of the XML import

results in a performance advantage. Therefore, in the programme code, this import is independent of the model set up. In addition, the import is simple, since no further adjustments to the data are necessary due to the structured structure of the XML file. The data type, such as string or integer, is adopted and thus saved in the simulation software. The internal table in the simulation software matches the structure of the XML file, i.e. all attributes correspond to columns and the single objects are listed line by line.

The model setup is also based on an independent method. For this purpose, the internal table generated by the import is run through step by step. Each line corresponds to an object of the model, which means that a simulation module is created here in each instance. A case distinction is performed in the first step, depending on which object type is given. For example, a different generation process is carried out for a machine object type than for a conveyor object. In a second step, a section of the model is then generated for each object. First, the position data is used to place the object. This data is scaled to the size of the simulation model area. Then the specific data is added to the generated module. Again, this differs depending on the type of object.

When the method is completed, a rudimentary model is generated. This completes the automated part of the method. For further automated generation steps, user input is required. For example, an additional method was set up that implements predecessor and successor relationships in the model based on the production plan. However, this requires a production plan, which cannot be determined by the scan. In order to design the process as efficiently as possible, predefined tables are stored in the model, which can be filled either with minimal effort by experts of the production system or by a defined export, e.g. from a MES. These input options are described in the further explanations. Figure 9.9 summarises the process.

The import is realized in the Plant Simulation software. An internal program function is used here to transmit the internal table with the data from the XML file. Such predefined functions are often found in other simulation software, which again underlines the advantage of using XML. The code for generating the model is also published and can be found in [20].

As a result, the rudimentary model is set up. Figure 9.10 shows an example of the result of a first application. It should be noted at this point that the example only shows a section of the production system. The different stations which all have the same shape in this form as they comply with the standard simulation modules can be seen. However, they have different characteristics due to stored properties. A visual distinction follows after the graphic interface has been adapted. In addition, it can be seen in the example that workplaces have been created next to the stations. This represents an extension that has not been explained before. In this case, workplaces were added for the specific application case of a project partner in order to be able to represent working personnel as well. In addition, there is a "DigiTwin" block in the

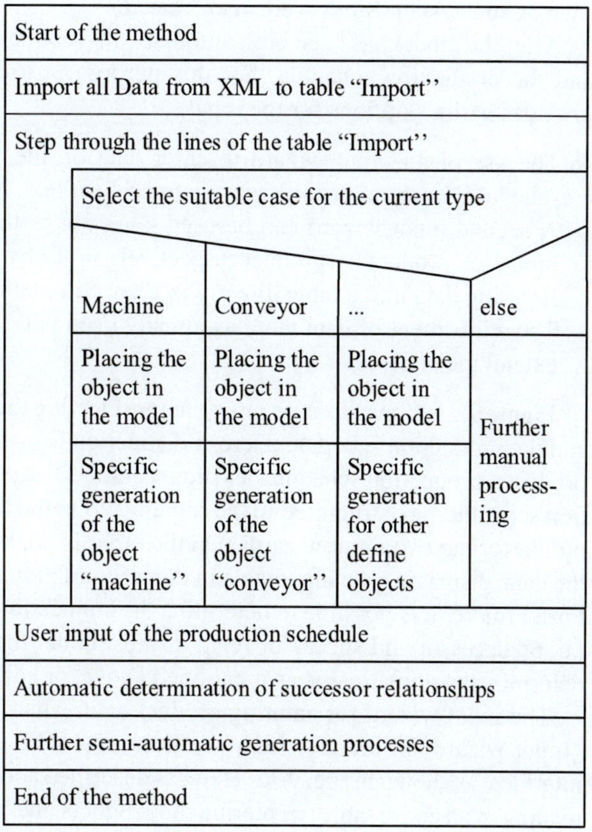

Fig. 9.9 Representation of the process with a Nassi-Shneiderman diagram [18]

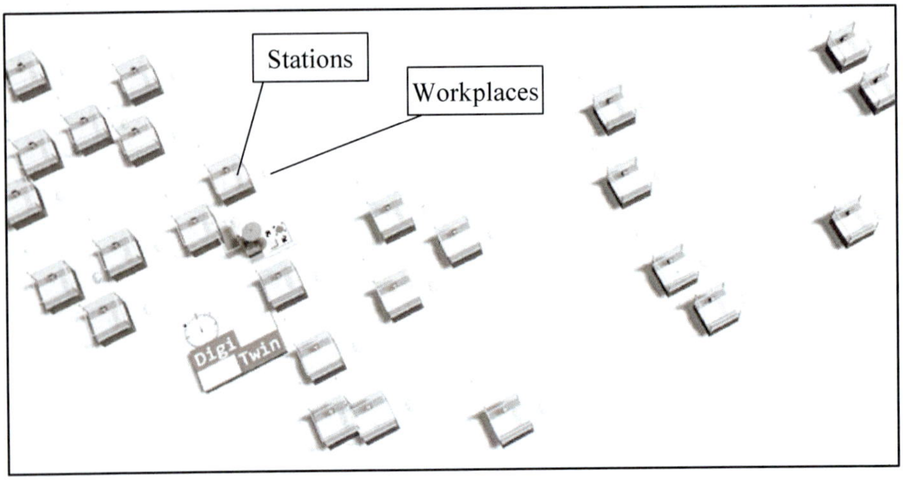

Fig. 9.10 Example for the functional model

model with which the user interface can be selected. A few process-relevant objects such as sources and sinks were also created.

After the model has been built automatically, the next step is to enter the products and the production schedule. For this purpose, a user interface was created that provides various options for the input:

- The use of an input wizard to enter data on the products and the production schedule. Here, an export from a spreadsheet (e.g. MS Excel) can also be done.
- A second input wizard can be used here, but in this case, it allows single-step input, i.e. product by product respectively order by order.
- Entering data into a table directly in Plant Simulation.
- The adjustment of data that has already been entered, for example, to update or extend the entries.

Figure 9.11 shows the user interface for selecting the type of entering the products and the production schedule. Here, a distinction is made between the three sections: products, production schedule, and connections. The points "products" and "production schedule" are structured in the same way. On the one hand, there is the selection option for the two input wizards. On the other hand, there is the option of adjusting the data, displaying the input table in Plant Simulation and deleting the data via reset. Furthermore, it is possible to determine the connections between the single stations, i.e. predecessor and successor relationships, under the point "connections". This is determined automatically, as mentioned above, or adjusted via a table.

The input wizard for entering product data, which can be activated by selecting "Input wizard" in the user interface from Fig. 9.11, is controlled via another user interface, as shown in Fig. 9.12. Here, as described above, there is the option of either reading in an entire table or entering the products one by one with a single-step input. For input via table, the formatting rule is displayed within the box. This formatting

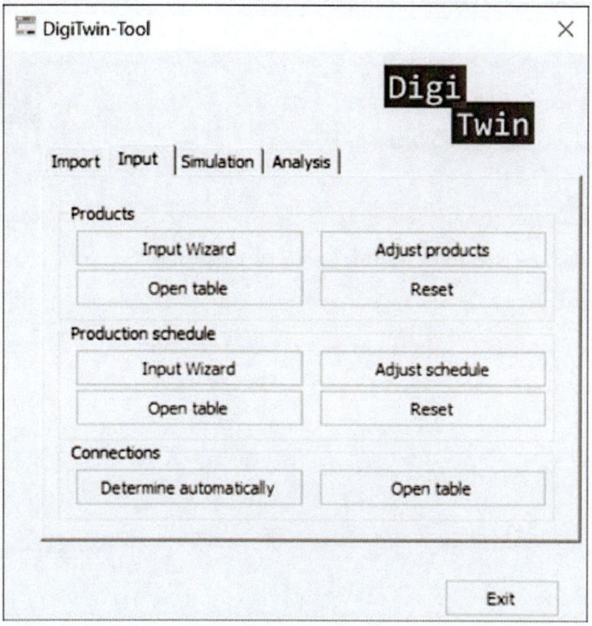

Fig. 9.11 User interface for entering the input

is then processed by a method so that all products are correctly integrated into the application. According to the structure of the data from the ERM in Fig. 9.5, the products are stored in a table. Further, the work plans and the work steps are stored in other tables and the corresponding links are set as defined in the ERM. This means that single work steps are assigned to the products via the work plans, which in turn are assigned to the stations where the products are to be processed. This results in an implicit process description when the products are entered, derived from the ontology described at the beginning. After all the data has been entered into the tables by the method, the products are then created automatically. The same procedure is also used for the single-step input. Here, the tables are filled with the data in the same way. However, after each input, the products are generated separately as described and the corresponding links between the tables are set.

So far, the described procedure has been implemented by exporting the product data from an MES. This export is usually not available in the required format. Therefore, the MES export has to be converted into the required form with the help of an MS Excel macro. However, it is conceivable at this point that an MES export is generated in the appropriate format or that an interface is set up between Plant Simulation and a specific MES application with which the data is transferred in the appropriate formats. In practice, different MESs are applied or the data may have to be extracted from ERP systems, so there is no generic solution available so far. As a result, there is manual effort for data pre-processing at this point.

The procedure for entering the production schedule is similar. Here, the input wizard also distinguishes between input via a table and the single-step procedure.

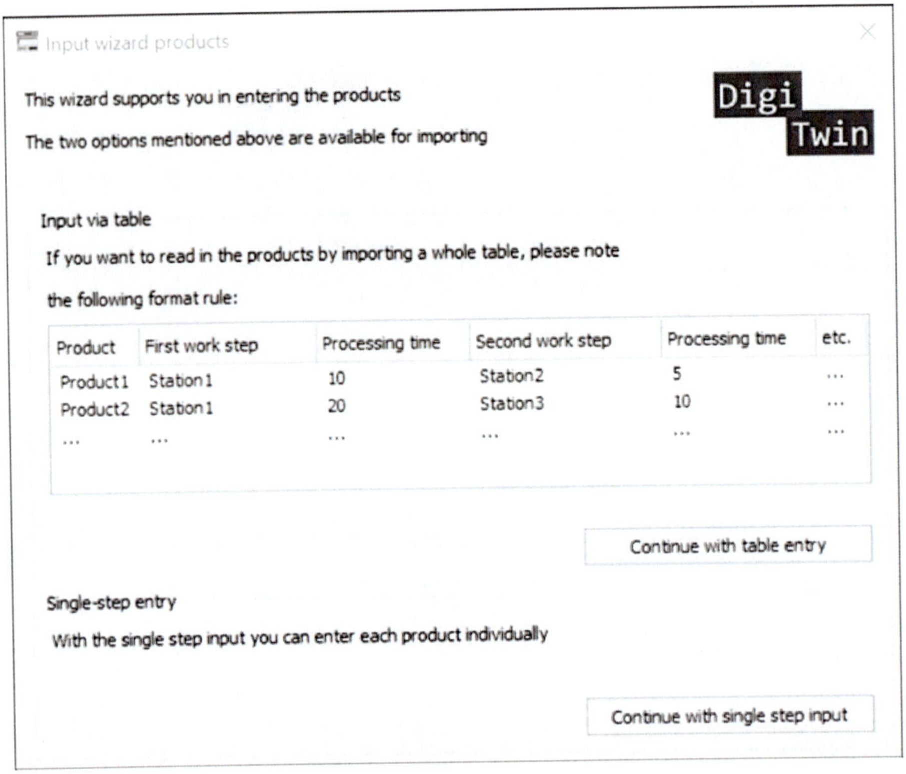

Fig. 9.12 User interface of input wizard (product data)

For the sake of completeness, this input is shown in Fig. 9.13. The corresponding formatting instructions can be found in the figure. The rest of the procedure is the same as for the products. Again, the data is written into corresponding tables in Plant Simulation and the links are defined according to the ERM from Fig. 9.5. Finally, the production schedule is automatically stored in the model. The difficulty with a MES export was also solved here with a MS Excel macro. A generic solution is currently not available, so that further manual effort is required for data pre-processing at this point.

With this specification, a rudimentary but executable model is set up according to Fig. 9.5. As already indicated above, some extensions were built into the first application. This refers in particular to the integration of workers and the insertion of intermediate buffers. These extensions are created automatically because they have been pre-programmed. The integration of further parameters for the simulation, as defined in Chap. 2 of this book, is also feasible. For this, either the creation process must be generically extended or the rudimentary model created here must be manually modified to include the required parameters. In the end, the functional model is built with this. However, in order to create a comprehensible and true-to-scale model, a graphical interface must be added. This is explained in the following section.

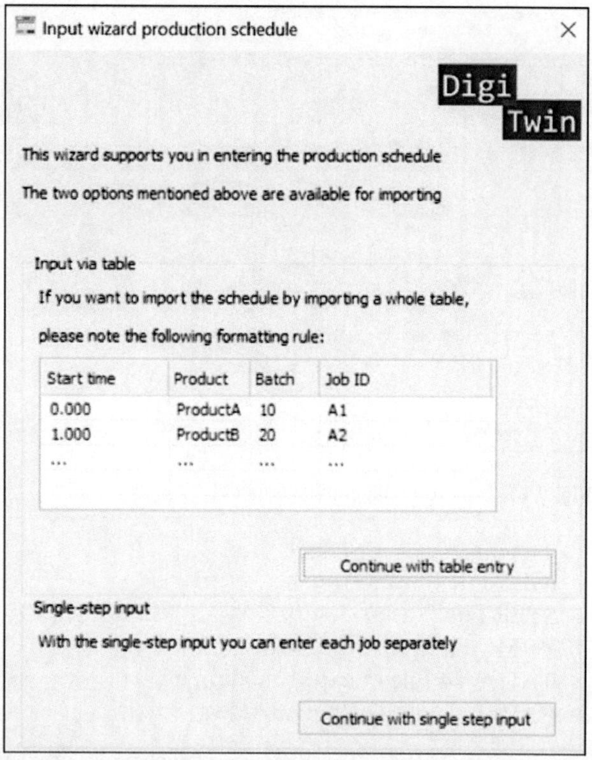

Fig. 9.13 User interface of input wizard (schedule data)

9.4.2 Design of the Graphical Model

The functional model provides the methodology for the simulation model so that the processes of the real system are correctly represented in the simulation programme. This means that the production process can be mapped to the model in terms of process times, status, transport times, malfunctions, etc. Since analyses are usually performed with the model, the findings of which can be applied to the real system, it is reasonable that the real layout of the production hall can also be reproduced in the model. This enables people who have no expertise in the operation of simulation software to understand the processes in the simulation model. Accordingly, this aspect is decisive for the acceptance of the simulation results.

As described above and in Chap. 8 of this book, the graphic model is first created as a CAD model as a result of the object recognition. Since the CAD software has various export options and Plant Simulation is able to load these exports, the transfer of the graphic model is simple. Various formats, such as:

- JT-File (*.jt),
- Parasolid-Text file (*.x_t; *.xmt_txt),
- Parasolid- Binary file (*x_b; *.xmt_bin),
- SolidEdge-File (*asm; *.par; *.psm),

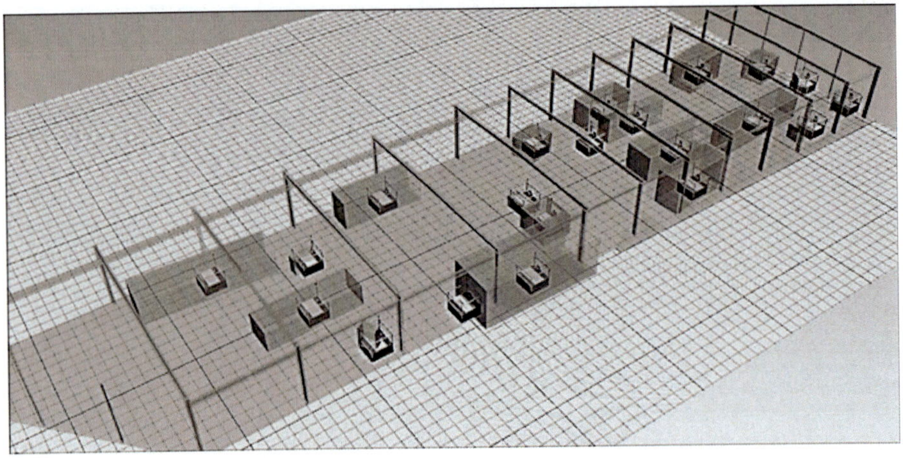

Fig. 9.14 Importing the graphical model

- IGES-File (*.iges; *.igs),
- STEP-File (*.step; *.stp),
- VRML-File (*.vrml; *.wrl),
- CATIA V4 File (*.exp; *.model), or
- CAD- Layout files (*.dgn; *.dwg; *.dxf),

can be used here. The decisive aspect here is that the same reference coordinate system is used for the graphic model as it was used for the export of the objects (see Sect. 9.4.1). This ensures that the graphical model corresponds to the functional model and that the graphical objects are placed exactly on the previously created objects in Plant Simulation. The graphic model is simply loaded into the software Plant Simulation and then aligned to the zero point, just as the previously created objects are also aligned to this zero point. By hiding the functional model, only the graphical model is visible to the user, while the functionality runs in the background. Figure 9.14 shows how the graphic model, which is marked in yellow in the figure, is placed over the existing objects. However, as described, this is only an intermediate step, as the objects of the functional model are then hidden.

The model construction is completed with the insertion of the graphical model and the hiding of the functional model. In the next section, the findings are summarized and the results of the model construction are presented.

9.5 Results and Summary

The ontology described in Sect. 9.2 is the basis for the further development of a semi-automatic model setup for simulations of production systems. With this approach, it is thus possible to develop a generic method with which any production system can be transferred into a simulation model.

For this purpose, it was shown how the data of a rudimentary model can be transferred into a database model so that the systematic description of production systems can be stored in a relational database. This procedure allows the required information to be stored in the simulation software and to be accessed in the model construction. The relational database schema also ensures the possibility of any required scaling and adaptation of the model.

In the descriptions, a rudimentary model consisting of stations, products, work plans, work steps and jobs has been intentionally introduced, as this represents a basis for mapping any production system to a model. This basis can be extended as required by the parameters introduced in Chap. 2 of this book. In the use case presented below, the rudimentary model was extended by the following points:

- In addition to the stations, buffers were also integrated into the model. These are used as intermediate storage before or after the processing operations. The buffers were integrated in order to correctly map the idle times in the model and thus to be able to observe the throughput time of the various products. The resulting system behaviour is needed for later planning, such as inventory management or scheduling. For the sequence of the application in the simulation software, the buffers are assigned to single stations, so that the logic of the original model remains the same and the sequence is only extended by the idle times. An extension of the ERM in Fig. 9.5, and therefore the database model, is not necessary in this case.
- In order to represent workers, workplaces were added to the stations. Next, workers were appointed at the workplaces so that a machining process can only be performed if a person is present. A possible extension at this point is to map the work efficiency. This could show the influence of the worker on the production process. However, this extension has not yet been implemented. The single workers are stored in a separate table in the simulation software, so that the ERM from Fig. 9.5 was extended by this table.

The purpose of listing these points is to explain which extensions are the basis of the following model. In addition, this serves as an example of which extensions are possible or necessary at this point in order to turn the rudimentary model into a simulation with which planning processes can be executed in a useful way. The model setup was presented after the database model was introduced. Here it is described that the model setup consists of two steps, creating a functional model on the one hand and a graphical model on the other. Both models are based on a factory hall scan and a downstream object recognition as described in the previous chapters of this book. The functional model is achieved from the CAD model, which is the result of object recognition, using a SolidWorks export, an XML interface and an import function in the simulation software Plant Simulation. The graphical model is implemented using common data formats. When both models are inserted into the simulation software, the functional model is hidden, leaving a detailed graphical interface for the user, but with the simulation functionality running in the background.

The result of this model construction is shown in Fig. 9.15. Here it can be seen that the model is about as accurate and detailed as a CAD model of the factory hall.

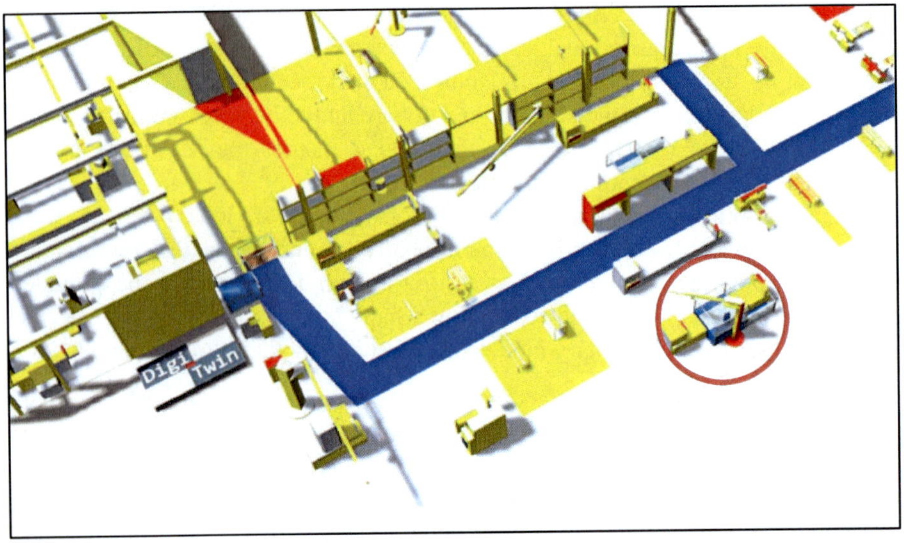

Fig. 9.15 Illustration of the final model

For illustration purposes, one station of the functional model was not hidden, so that it can be seen in the area outlined in red that a standard component of the simulation software is also located where the machine of the graphic model is.

It has already been mentioned that setting up a simulation model of a production system is time-consuming and requires expertise in the corresponding simulation software as well as in the operation of production systems. The approach presented here provides an efficient and highly automated solution. The automated part of the method includes:

- Preparation of the scan of the production hall into a point cloud,
- object recognition based on the point cloud to identify the relevant objects of the production system,
- export of the functional objects into an XML interface,
- import of the objects into the simulation software,
- creation of the objects in the simulation model,
- creation of products after data input,
- integration of the production schedule after data input,
- creation of connections between the various stations, and
- pre-programmed, simple analyses of the production system.

Nevertheless, there is manual effort that is also necessary for this procedure. However, this is limited to a large extent, so that an efficient procedure could be developed overall in comparison to the conventional procedure. The manual effort consists of:

- Scanning the production hall,
- data acquisition and pre-processing for product data,

- data acquisition and pre-processing for production schedule,
- extension of the simulation model to include company-specific properties, as well as
- target definition for simulation analysis and interpretation of simulation results.

This manual part is removed from the end user by a service concept, which is described in more detail in Chap. 11 of this book, so that no expertise is needed for the model setup and for the analyses. It is also possible to expand this approach to include the development of interfaces to MES, so that data acquisition and pre-processing is further simplified. This further development is made possible by the approach with its data-technical structuring without major adjustments. The time required is reduced to approximately two working weeks. Compared to the conventional procedure (for example in [1]) for model creation, this corresponds to a time reduction of 81 percent.

References

1. Fritzsche P, Abel D, Rudolph T (2017) Simulationsergebnis und das Ergebnis der Simulationsstudie in Form der resultierenden Handlungsempfehlungen – Fallbeispiel einer Simulationsstudie bei der Rudolph Logistik Gruppe. In: Wenzel S, Peter T (eds) Simulation in Produktion und Logistik 2017. Kassel University Press, Kassel, pp 219–228
2. Stobrawa S, Wilmsmeier S, Denkena B, Dittrich MA (2019) Automatic generation of model sets for simulation-based validation of new production planning and control methods. In: Proceedings of the 9th congress of the German Academic Association for production technology (WGP), Production at the leading edge of technology, Hamburg, pp 643–652
3. Kikolski M (2017) Study of production scenarios with the use of simulation models. In: 7th international conference on engineering, project, and production management, procedia engineering vol 182, pp 321–328
4. Lödding H (2016) Verfahren der Fertigungssteuerung - Grundlagen, Beschreibung, Konfiguration, 3rd edn. Springer-Verlag, Berlin, Heidelberg
5. Bergmann S, Straßburger S (2015) On the use of the core manufacturing simulation data (CMSD) standard: experiences and recommendations. Fall Simulation Interoperability Workshop, Orlando
6. Pellegrini T, Blumauer A (2006) Semantic Web – Wege zur vernetzten Wissensgesellschaft. Springer Verlag, Berlin, pp 9–27
7. Chandrasekaran B, Josephson JR, Benjamins VR (1999) What are ontologies, and why do we need them? IEEE Intell Syst January/February 1999:20–26
8. Simulation Interoperability Standards Organization (2010) Standard for: core manufacturing simulation data—UML Model. Core Manufacturing Simulation Data Product Development Group. SISO-STD-008- 2010. http://www.sisostds.org/DigitalLibrary.aspx?Command=Core_Download&EntryId=31457
9. Akerkar R (2009) Foundations of the semantic web XML, RDF & Ontology. Alpha Science International Limited, Oxford
10. Block C, Lins D, Kuhlenkötter B (2018) Approach for a simulation-based and event-driven production planning and control in decentralized manufacturing execution systems. Procedia CIRP 72:1351–1356
11. Ellwein C, Elser A, Vollmer M, Riedel O (2020) Connected production planning and control systems—implementation and the optimization process for subcontracting. Procedia CIRP 88:191–196

12. Lorenzen LE (2012) Entwicklung einer Methode zur gentelligenten Arbeitsplanung. Leibniz Universität Hannover, Dr.-Ing. Diss
13. Fournier J (2011) Model building with Core manufacturing simulation data. Proc Winter Simul Conf 2011:2214–2222
14. Dangelmaier W, Gausemeier J (2019) Intelligente Arbeitsvorbereitung auf Basis virtueller Werkzeugmaschinen. Springer Vieweg, Berlin, Heidelberg
15. Denkena B, Dittrich MA, Stamm S, Wichmann M, Wilmsmeier S (2021) Gentelligent processes in biologically inspired manufacturing. CIRP J Manuf Sci Technol 32(2021):1–15
16. Bergmann S, Straßburg S (2010) Challenges for the automatic generation of simulation models for production systems. Int Comput Simul Conf (SCS). Ottawa 2010:11–14
17. Denkena B, Stobrawa S, Sommer M, Stjepandic J, von Soden M (2020) Produktionsplanung mit dem digitalen Zwilling - Effiziente Erstellung eines digitalen Zwillings zur Verwendung im Produktionsmanagement. wt Werkstattstechnik online 110(10), pp 661–665
18. Denkena B, Dittrich MA, Stobrawa S, Stjepandic J (2020) Efficient generation of a digital twin using object detection for data acquisition and XML-interface for model creation. In: 53rd CIRP conference on manufacturing systems, proceedia CIRP 93, pp 274–279
19. Stobrawa S Solidworks export-XML IFW DigiTwin. https://doi.org/10.5281/zenodo.3629369 downloadable under https://zenodo.org/badge/latestdoi/236736366
20. Stobrawa S Plant simulation import & modell generation IFW DigiTwin. https://doi.org/10.5281/zenodo.3618211 downloadable under https://zenodo.org/badge/latestdoi/23535622

Chapter 10
The Commercialization of Digital Twin by an Extension of a Business Ecosystem

Josip Stjepandić, Markus Sommer, and Sebastian Stobrawa

10.1 Introduction

Industrial products, production systems and product services are designed and produced, applying distributed engineering and production approaches which are supported by product lifecycle management (PLM) systems [1]. These systems integrate business processes and systems in an extended enterprise, understood as a loosely coupled network of companies that thrive on effective partner management and continuous self-organization as well as learning. Original Equipment Manufacturers (OEM) are continuously faced with the challenge to provide innovations, while reducing their time to market, cutting costs and improving product quality [2]. To be competitive, OEMs steadily focus on their core competencies and outsource the production of components, various subsystems, modules and services [3]. To enter markets worldwide, manufacturing companies and especially OEMs have established globally distributed supplier networks which significantly contribute to their market success [4]. Proper supplier selection and integration followed by effective supply chain management are therefore crucial for OEMs' success [5]. For the supply chain to operate properly, all parties need to play their roles and learn continuously, in order to further develop the supply chain processes and make them more flexible.

J. Stjepandić (✉)
PROSTEP AG, Dolivostr. 11, 64293 Darmstadt, Germany
e-mail: josip.stjepandic@prostep.com

M. Sommer
isb innovative software businesses GmbH, Otto-Lilienthal-Strasse 2, 88046 Friedrichshafen, Germany
e-mail: josip.stjepandic@prostep.com

S. Stobrawa
Institut Für Fertigungstechnik Und Werkzeugmaschinen, Leibniz Universität Hannover, An der Universität 2, 30823 Garbsen, Germany
e-mail: stobrawa@ifw.uni-hannover.de

J. Stjepandić et al. (eds.), *DigiTwin: An Approach for Production Process Optimization in a Built Environment*, Springer Series in Advanced Manufacturing,
https://doi.org/10.1007/978-3-030-77539-1_10

A complex, dynamic, loosely coupled and self-adaptive system made up of these parties represents the supply network [6]. Due to the necessity to continuously coordinate and manage intense flows of material, product and information, it can also be called an ecosystem [7]. Ecosystem management plays an exposed role within the OEM-supplier relationship. It should be noted, however, that suppliers and service providers also build their own ecosystems, establishing relationships with multiple partners [8].

Services have existed since the earliest history of human beings as outcome of labor organization. Similarly, to the term product design, the term service design was coined when the relevance of services in economic activities became evident and the term blueprinting arose in order to illustrate the activity of designing and codifying the sequence of actions that are part of a service performance within almost all market participants [9].

Service underlies similar phases in its life cycle as a product and can be structured in a similar manner. Services can be analyzed focusing on the organization of business operations and describing them as processes. From a similar perspective, services are characterized as experiences that happen over time and that need to be organized through a sequence of interactions between service providers and customers. Service design has also been defined in relation to the coordination of the backstage of services, in other words, to the design of facilities, servers, equipment and other resources needed to produce services [9].

Two main tensions yield the definition of service design: the first tension results from understanding design either as a defined problem-solving activity or as an exploration of an open problem space involving different actors, including users (Fig. 10.1). The second tension concerns understanding services on the basis of how they differ from products or as an activity of value creation [10]. As a result of these tensions, design is defined either from an engineering perspective—keeping the distinction between products and services and interpreting design as a problem-solving activity—or from a design-for-services perspective, which looks at services as a value creation activity in an open-ended problem exploration involving different actors [10]. Using this classification, approach presented here can be clearly assigned to the area of Service Engineering: value creation for the well-defined problem of creating a Digital Twin (DT) [11].

OECD (Organisation for Economic Cooperation and Development) states that an ecosystem is a coherent set of entrepreneurial (potential as well as active) players, market parties, institutions (government, universities and watchdogs) and processes formally and informally influencing a local environment [12]. In recent years, ecosystems became popular among leading global companies [13].

In this chapter, essential steps for commercialization of DT are discussed. Section 10.2 presents main challenges of supplier networks, engineering collaboration and network-centric operations, followed by Sect. 10.3 on product and service platforms. Section 10.4 explains the main features of OpenDESC.com, followed by a description of use cases in Sect. 10.5. Planning of extension by enterprise architecture integration is discussed in Sect. 10.6, and adoption of extended services in

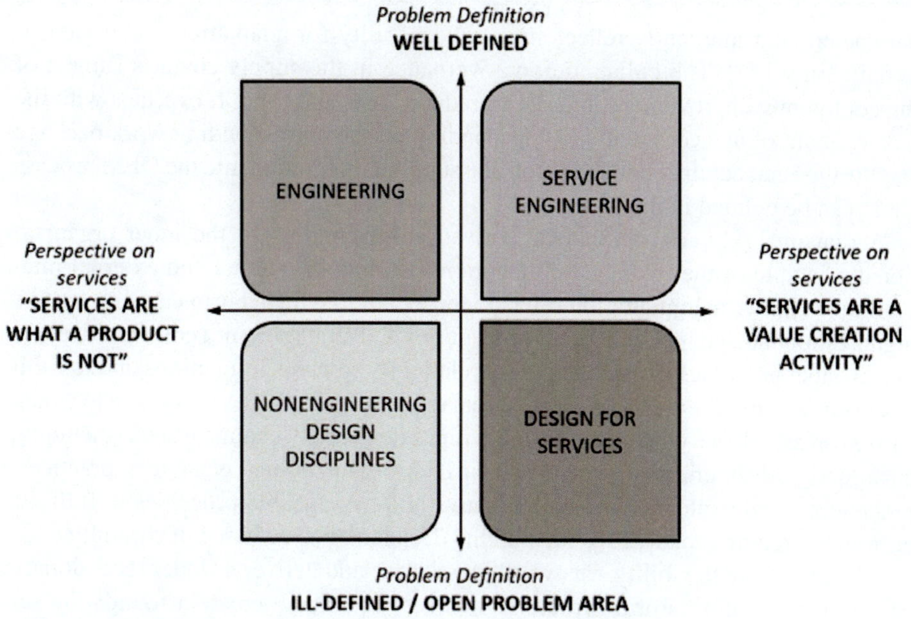

Fig. 10.1 Approaches to conceptualising service design, derived from [9]

Sect. 10.7. Adoption of Digital Twin offering is assessed in Sect. 10.8, following with discussion in Sect. 10.9. The chapter ends with a conclusion and outlook.

10.2 Supplier Networks, Engineering Collaboration and Network-Centric Operations

Industries like automotive, aerospace or shipbuilding with their globally extended enterprises offer typical examples for manufacturing supply networks. In these industries, the rise of outsourcing has led to a multi-tier supply network structure involving numerous partners around the globe [3, 14]. These networks demonstrate new tendencies within the production organization, such as global manufacture, strategic alliances, flexible production and mass customization [15]. A leading automotive OEM is present in almost each country, earns more than 150 billion US$ with more than 500,000 employees, and maintains a network with several thousand suppliers and service providers [3]. Alliances have emerged as a typical expression of business with the aim to share opportunities and risks [16].

Critical business processes run in such partnerships [17]. Thus, the term 'virtual enterprise' is coined comprising the 'core competence cells' [14]. Networks are becoming the grail for agility and adaptability [18]. Ascribing critical success factors against collaborative alliances raises multiple possible, and potentially measurable, parameters that inherently reflect strategic capability for adaptation and organizational agility [14]. The collaboration governance in the supply chain is subject of the continuous improvement in order to reduce costs and time. It overlaps with the management of an ecosystem [19]. Depending on the corresponding work package and the type and depth of collaboration, the supplier integration into the OEM process chain can be defined in many ways [3].

A dynamic role concept can be derived corresponding to the inner organizational principle of the two-level-collaboration with both internal and external integration [14]. Internal integration refers to the extent to which a firm can structure its organizational behaviors, practices, procedures, and strategies into collaborative and manageable processes to establish close relationship between functions and to fulfil its customers' requirements. While external integration describes the degree to which a firm can establish a close relationship with its key supply chain partners (including customers and suppliers) to structure their inter-organizational behaviors, practices, procedures, and strategies into collaborative and manageable processes to fulfil its customers' requirements [20]. Outsourcing decision must be based of carefully evaluated metrics like flexibility, trustworthiness, price and delivery. Under recessionary constraints a 'build the infrastructure' was recommended: create a foundation for integration to enable 'plug and play' independent and integrated applications with focus on infrastructure which will be rewarded in the upturn [14].

Looking from the broader view, for example, the area of defense, network-centric operations (NCO) promise increased reaction speed and improved quality of decision making made possible by greatly improved situational awareness and access to widely dispersed forces and weapons. The operating conditions of the operational environment are characterized by frequent changes in products, services, processes, organizations, markets, supply and distribution networks. NCO are characterized by the rapid acquisition, processing, and exchange of mission-essential information among decision makers at all command levels. This enables them to operate from the same, verified, situational and targeting knowledge bases at the resolution and the decision cycle time required at each level. Network communications is the service-based foundation for linking or networking systems to share information across geographic borders and dynamically reallocate resources based on operational needs [14].

With the rise of the network and the alliance, two distinctive risks rise too: relational risk and performance risk. Relational risk concerned with the probability partners would not comply with cooperation of the alliance, while performance risk referred to the probability that strategic goals of the project would not be achieved. In an alliance structuring process, the partners faced with two sets of risk: those regarding future states of nature and those regarding cooperation. While helping to focus on where risks could reside in an alliance, more effort is required to identify and manage these risks [21].

Exploring the nature of services as interactions between different actors allows to study services from a different perspective. The relevance of interaction aspects also relates to the time dimension, which qualifies services as processes rather than products. Another consequence of this perspective is that the value is co-created in

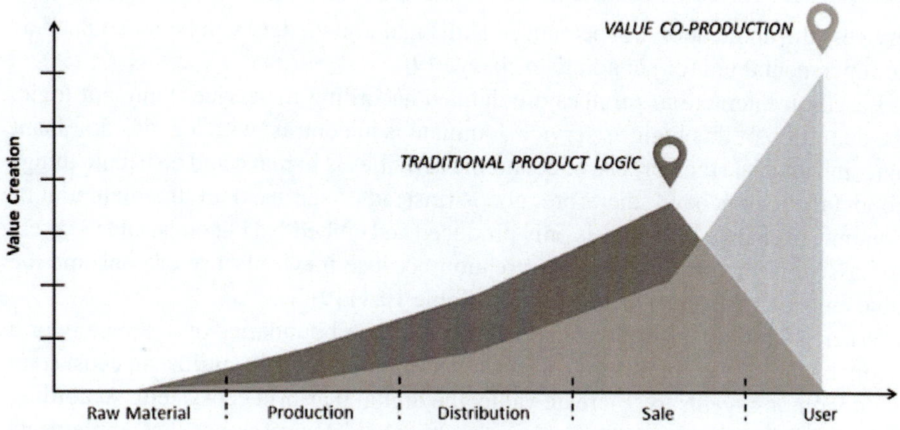

Fig. 10.2 Goods versus service value co-production, derived from [9]

this interaction, instead of just being passed from a producer to a consumer. Services are an activity of value creation that continues (and sometimes becomes more intense) after the point of sale, for example, after the moment in which a contract for the service is signed or after the material resources to produce value are transferred from the service provider to the customer. The point of sale is the beginning of a co-creation process based on a collaboration between customers and other actors (including the service provider) (Fig. 10.2) [9].

10.3 Shift from Product to Service Platform and Ecosystem

Product platforms are a collection of modules or parts that are common to a number of products [22]. A service platform is a grouping of related services that are similar in resource type and constitute a component of a continuum of care [23]. Products produce a single revenue stream, while platforms—which we define as intermediaries that connect two or more distinct groups of users and enable their direct interaction—can generate many. Indeed, a large number of the world's most valuable companies by market capitalization in 2015 were platform companies, including five of the top ten. Four steps were identified that can make the difference between effective transformations and failure [24]:

- Start with a defensible product and a critical mass of users
- Apply a hybrid business model focused on creating and sharing new value
- Drive rapid conversion to the new platform
- Identify and act on opportunities to deter competitive imitation.

Service as a value production process does not only apply to contemporary services (e.g. service platforms) but also could be an approach to interprete any kind of service and also any kind of product with the same logic. Alternatively, services (the application of specialized skills and knowledge) can be understand as the fundamental unit of economic exchange [9].

Based on interaction, services are defined according to service dominant logic. The definition of this logic as service dominant is in contrast with a goods dominant logic, in which (1) the purpose of economic activities is to make and distribute things (goods) that can be sold; therefore, goods (instead of services) are the main unit of economic exchange, (2) value is only produced and embedded in goods, and (3) users are passive in the process of value creation because they only use or consume the value embedded in a product or service (Table 10.1) [9].

Whereas traditional firms create value within the boundaries of a company or a supply chain, service platforms, in particular digital platforms, utilize an ecosystem of autonomous agents to co-create value: the digital platform ecosystem. According to this, a digital platform ecosystem comprises a platform owner that implements governance mechanisms to facilitate value-creating mechanisms on a digital platform between the platform owner and an ecosystem of autonomous complementors and consumers [25, 26].

From a structural point of view, four constituents can be observed in an ecosystem: the keystone organization (the platform leader), members of the ecosystem, a dominator among members and niche players. The keystone organization is the company that is capable to provide platforms that serve other members in developing their products and ensure their market survival. Although they represent a small part of the ecosystem, they account for a larger share of profit. The keystone organization governs the whole ecosystem in terms of influencing actions and development of all members and defining the ecosystem´s architectural design [7].

The direct rival to the keystone organization, also referred to as the "wannabe", is a member that competes for the governance of the ecosystem. If the "wannabe" tries to replace the keystone organization via vertical and horizontal integration with a perspective to govern the ecosystem, it is called a "dominator". Finally, niche players bring a large share of innovations into the ecosystem through their specialized functions and thus act as complementors. By focusing on a niche, they usually outsource

Table 10.1 Goods dominant logic versus service dominant logic, derived from [9]

	Goods dominant logic	Service dominant logic
Role of goods	Goods are end-products	Goods + embedded knowledge composed into the process of value creation
Role of customers	The customer is the recipient of goods	The customer is a co-producer of value
Value producer	Value is determined by the producer and embedded in goods	Value is perceived and determined by the customer
Firm/customer relation	Customers are passive	Customer is active, firms can only make a value proposition (through products/services)

some functions to other members, utilizing the platform of the keystone organization and products of their complementors [7].

From technological point of view, digital platforms are built on a modular architecture comprising a stable core and a flexible periphery, taking advantage of economies of scale and substitution [25]. For an ecosystem to be successful, timing in the development of components and complementary technology is crucial [7]. Through the modular infrastructure, the platform owner provides affordances that complementors can actualize based on individual innovation capabilities. On the other hand, complementors can interact with each other to utilize the generativity of the digital platform ecosystem [25].

Activities are discrete actions that determine how value is jointly created in an ecosystem. In a digital platform ecosystem, actions include the development of new applications or the provision of services, such as offering rides or listing new properties. Interdependencies of actors and products in an ecosystem can result in bottlenecks whose performance, costs, and scarcity might constrain the value proposition of an ecosystem. Complementary bottlenecks could lead to failure of the whole system because they could prevent the focal firm from showing its performance to the customer. Research on digital platforms has suggested that platforms act as bottlenecks to control and limit interactions in an ecosystem [26]. In following, it will be explained how an ecosystem were built and how an innovative extension is being adopted by a long running ecosystem.

10.4 The Rise of OpenDESC.Com

Central component of a digital ecosystem is a software platform that is built to support the value creation process and manage complex socio-technical processes [27]. Figure 10.3 depicts the operating ecosystem based on OpenDESC.com service portal which serves as a hub for data communication in the automotive industry and

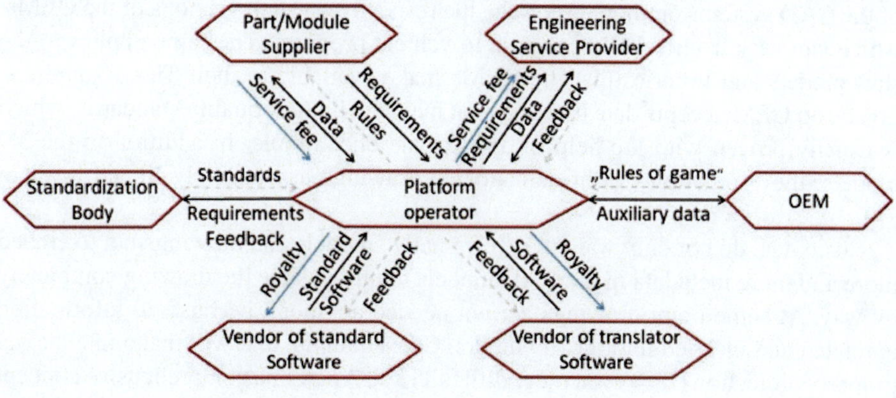

Fig. 10.3 Structure of OpenDESC.Com ecosystem [7]

enables acquisition and dissemination of data regarding logistics needs. It is a cloud-based service that ensures that data reach stakeholders and customers in time, in a secure manner and with the required level of quality [7].

Apart from the parties in engineering collaboration, there are few IT related stakeholders: platform operator (as operator of the ecosystem), software vendors and standardization bodies. Arrows illustrate different types of mutual relationships [7] such as requirements, rules, data, standards, provision of software, payment and feedback. The structure and the relationships in this ecosystem are subject to continuous change and evolution, resulting from changing trends in the market, organization, methods and technology. The ecosystem is thus a dynamic, self-adapting system which follows the market dynamics and societal change [28]. In times of crisis, as is often the case in the automotive industry, the demand from many suppliers drops subsequently. However, new technological trends (e.g., electro mobility) emerge, contributing to changes in the ecosystem and the dynamics supported by OpenDESC.Com.

Established in 1998, OpenDESC.Com has evolved to become a holistic service for enterprise-wide data logistics, which includes both the conversion and the transfer of the converted data to the partner [29]. Via an intuitive service portal, users can provide their CAD data, as well as other product-related documents, automatically for exchange and import the data and documents provided for them into their PLM systems. Before transferring the CAD data, they can be converted into the format required by the receiver considering the OEM-specific requirements.

The reason for founding the Data Exchange Service Center (DESC) was the need of the automotive suppliers to convert their CAD data into the format of the OEM's favorite system CATIA V4, without having to purchase expensive CAD workstations themselves. Two years later, when no one was yet thinking about cloud or web-based services, OpenDESC.Com was established, the first web portal for data conversion. It allows customers to upload their data for conversion without installing additional software and to download the converted data again [29].

The requirements of automotive manufacturers for data quality and data security in cross-company communication have risen continuously over the past decades. As a rule, system suppliers today must deliver native data. Native means in the formats of the CAD systems or, more precisely, the respective system versions of the OEMs, which can vary from vehicle program to vehicle program. The same applies to the start models and libraries that the OEMs make available to their Tier 1 suppliers. Today, no OEM accepts data that does not meet the highest quality standards, which is usually proven with the help of appropriate check tools. In addition to the 3D models, they sometimes even demand 2D drawings associatively linked to them [30].

Customers do not only want to receive native models and drawings but to embed more and more metadata in the CAD models to quite waive the drawing completely by way. At some point drawings are not needed anymore, because all information is related and inserted into the 3D model. Consequently, that will make intellectual property protection (IPP) even more difficult [31]. Apart of a comprehensive concept of intellectual property management, an appropriate CAD data export filter must be used for such a scenario [32].

Data transfer has also become more complex over the years. While automotive suppliers used to be able to send their data as individual files via secure file transfer, OEMs are increasingly demanding that they place complete assemblies in their PLM systems via a portal or a special client. In line with market requirements, the range of services offered by OpenDEC.com was expanded to include other services such as data transfer. It is playing an increasingly important role for automotive suppliers due to the growing importance of data security and the growing complexity of communication processes. The ability to encrypt data on the portal during the exchange has become crucial for secure data transfer [17].

Outsourcing data conversion and data transfer not only saves suppliers the cost of purchasing CAD systems and data exchange tools, maintaining and continuously adapting the systems to the changing specifications of OEMs, and qualifying the personnel who have to administer and run the systems. It also improves their productivity by enabling them to design and optimize their internal processes more effi-ciently in a homogeneous system landscape, regardless of customer specifications. Using OpenDESC.Com also guarantees consistently high data quality, without which no supplier would receive orders today, and maximum data security during data transfer. Logging and archiving of orders also ensure maximum transparency. Another advantage for globally active suppliers that should not be underestimated is that they can use the data logistics service virtually around the clock.

10.5 Use Cases of OpenDESC.Com

From the customers' point of view, OpenDESC.Com is a modular data logistics platform that supports different use cases from data conversion to migration and data transfer (Fig. 10.4) [17]. It primarily addresses customers who want to buy extended services in this area instead of build-up of their own infrastructure. It enables suppliers to optimize their internal development environment according to their specific needs

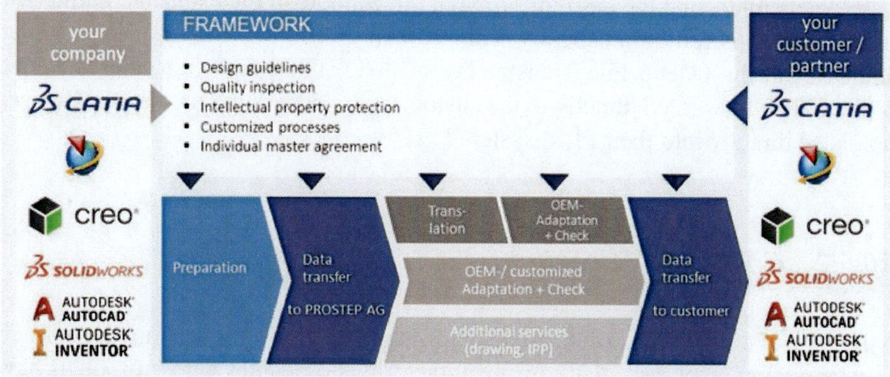

Fig. 10.4 Block schema of OpenDESC.Com [14]

instead of replicating their customers' respective environments. They also do not need to maintain a heterogeneous CAD system landscape and train employees to use different CAD systems in order to convert their development data into the native customer formats (Fig. 10.4). OpenDESC.Com combines optimal intellectual property protection during data conversion with maximum security when transferring data to customers and clients [33].

OpenDEC.com is currently used by numerous customers around the word. The basis of its success is the data logistics experts' deep understanding of the OEMs' requirements. Thanks to their good knowledge of OEM-specific IT environments, they are able to provide customers with remote support for all issues relating to the installation of OEM-specific software tools, ensuring that their system landscape is always at the release level specified by the manufacturer. This high flexibility is crucial for the emergence of an ecosystem [34].

10.5.1 CAD Data Translation

Data conversion into all common CAD formats is a mainstay of the data logistics service, with various options available to the customers. The most cost-effective is automated conversion without quality control, which usually takes no longer than 24 h. It is popular with customers with high-quality source data or in early development phases, where speed is crucial. For customers with higher quality requirements, the service platform ensures quality control of 3D models and 2D drawings, adapts customer's data to OEM-specific specifications and corrects any errors in the target system during conversion. To facilitate the further use of the data, OpenDESC.Com, offers the conversion of parametric and feature-based CAD models as an additional option [35]. If the method presented in this book is understood as a CAD translation from a point cloud format (Fig. 10.4, on the left-hand side) to a parametric CAD format, then it counts to this type of service. From the user perspective, the conversion process requires minimal interaction. The users only need to select the files to be sent, the recipient and the desired format in the window of the exchange portal—all other operations run automatically without their assistance. Based on pre-defined addresses in the Odette File Transfer Protocol (OFTP) connection, the converted files are then forwarded directly to the customer or—depending on the OEM—even uploaded directly into their PLM system [17].

10.5.2 CAD Data Migration

Migration of existing CAD data is one of the major challenges when moving to a new system environment, especially for manufacturing companies whose products have long lifecycles and are regularly modified and improved during lifetime. Any small error is being multiplied within thousands of models which causes a non-predictable rework. Therefore, many customers draw on the long-standing data conversion know-how. In addition to data conversion, the service includes assessment on preparing the migration and defining the optimal migration strategy: From a so called big-bang migration (translation of bulk data within a short period) with several terabytes of data, to an incremental migration, to an on-request migration where only the data that is just needed is converted [36]. The challenge is how to convert large amounts of data in the shortest possible time. For this purpose, special adapters and pipelines, test routines as well as automatic repair processes for any migration process (Fig. 10.4) are created and key performance indicators (KPI) for effective success control are defined [37]. Clearly defined processes guarantee the success of the migration.

10.5.3 Data Transfer

Few providers offer data conversion and organization of data exchange with OEM and Tier 1 supplier from a single source. For this purpose, the highly secure data exchange platform was integrated into OpenDESC.Com, enabling customers to use the software as a service without having to install it. The data transfer service relieves automotive suppliers of all the tasks associated with data exchange. All they need to do is specify who is to receive which data; the service platform takes responsibility for ensuring that it reaches the recipient on time and in the right way. The exchange is handled either automatically via OFTP or OFTP2, via the OpenDESC.Com web portal or manually via the automotive manufacturers' supplier portals. For a seamless process, the database with the contact persons on the recipient side must be filled and the corresponding exchange profiles created. Personalized access ensures maximum protection of intellectual property. The data exchange solution logs all exchange processes so that customers can prove at any time which clients or suppliers received which data and when [17].

10.5.4 Intellectual Property Protection

The requirements of the respective process chain (design in context, digital mock-up etc.) yields the corresponding IPP measures. The data exchange agreements within the supply chain and the ecosystem as well differ with respect to the content and purpose of CAD data to be delivered, i.e. the degree of "parameterization" and the data exchange frequency. At this step, dedicated knowledge portions from the CAD data must be removed very precisely and context-sensitively to satisfy the different data exchange agreements. The requirement is often to deliver minimal knowledge with the CAD data, but to satisfy the data exchange and data quality thresholds. Depending on the use case there is a strong need to define some IPP rules for controlling the IPP process in a predefined way [31]. This service is provided by an IPP software which works similarly to an additional intelligent CAD workbench with a comprehensive filter function.

10.5.5 Portal Services

Some automotive manufacturers maintain portal solutions to give partners and suppliers direct access to their PLM systems and data. This means that their partners must be familiar with the respective OEM environments to search and find the data. The portal services are designed to act as intermediary and free them from this necessity. OpenDESC.Com meets these high requirements in terms of system know-how and available IT infrastructure. As a Certified Data Service Provider, the staff is authorized to handle all data communication between the OEM sites and its suppliers. This includes the ability to export data from the OEM's PLM systems for the suppliers, to convert it, to derive manufacturer-compliant drawings and to put data back into the OEMs systems. The use of the data transfer service contributes significantly to a fast and flexible integration between OEMs and suppliers and it also facilitates the unbundling of partner relationships if necessary [17].

10.6 Planning of Extension by Enterprise Architecture Integration

Coming from Systems Engineering, one of the most used methods for large change projects is Zachman Framework [38], composed by a high-level architecture of the platform on the perspective of the following six layers (Table 10.2).

Scope: Collaboration with OpenDESC.Com is an interesting alternative to cut costs and facilitate making the exchange processes uniform and ensures a higher level of reliability, without having to invest in a complete corporate infrastructure [39]. As described in previous chapters, generation of Digital Twin can be composed as an additional service offering which either entirely or partially consists of singular steps described in Sect. 3.5.

Business Model: Business entities are allowed to process orders between development sites. The service offering is governed by a Service Level Agreement (SLA) that covers processes like on-boarding new partners, transmitting and receiving data as well as support levels and decommissioning of registered users [38]. Having this in place, connections can be setup between engineering parties based on mechanisms and networks or by means of engineering portals [40]. Each party in the collaboration is provided with a platform on pay-by-use basis.

System Model: The system (logical) model is built of users and organizations, the actual offering and jobs (service orders) performed between users granted to use the service. Auxiliary infrastructure belonging to customers and OEMs are docked to

Table 10.2 High-Level Zachman Framework for OpenDESC.Com, adapted from [38]

	What	How	Where	Who	When	Why
Scope	Engineering product data provided by a customer or its partner are submitted according to specific constraints	Translation and exchange are thereby offered as cloud service	Global: All engineering sites of the involved parties are affected	OEMs and tier-X system suppliers of a customer	At each design change from an involved party, at beginning of planning activities	Achieve enterprise cross-collaboration without need for an infrastructure on premise
Enterprise model	**Entity** = Business Organizations placing Order having Engineering Data and Tracing collaboration activities based on Report **Relationship** = Contract (SLA)	**Process** = Change Management, On-boarding, Data Transmission, Data Reception, Support	**Node** = Business Locations (mostly engineering) **Link** = Connection	**People** = Sales, Design Department, Service Provide **Work** = Service Level Agreement, Process Order	On Data Submission, On Data reception	Enable engineering organizations to realize fast, easier and reliable communication demand
System model (Logical)	**Entity** = Electronic DataSets contained in a Job submitted to Users in Organizations having a contractual Agreement **Relationship** =Person in Organization	**Process** = Set Up, Export Data, Import Data **I/O** = Web All parties must be registered to the platform and enabled to use services.	**Node** = Business Unit **Link** = Web Portal, OFTP2	**People** = Sales Person, Designer, Process Planner, Administrator, Sender, Receiver **Work** = Engineering Design, Process Planning, Send, Receive Data, Manage Services	On request	A party may be requested to provide, on a given date, data in a particular configuration to a customer.

(continued)

Table 10.2 (continued)

	What	How	Where	Who	When	Why
Technology model (Physical)	Model based Physical Data models with diagrams of the technology architecture, control structures, definitions and descriptions are realized using UML techniques. Users are interfacing directly with the service using (1) a web front end or (2) by means of an application programming interface for machine-to-machine communication. The UI is thereby described using graphical mock-ups					
Detailed representations	By using templates, each customer of OpenDESC.Com is easily set up as a node with its specific characteristics based thereby. Each node is logically separated from other customers in order to fulfill stringent security requirements. The operation of a node is conducted over an administration and operation interface, which allows an update of the node database					
Functioning enterprise	Each authorized user can create communication relationships, define rules for exchange, timing, quantity, desired data quality, level of detail and protection of intellectual property, add required metadata, distribute data sets to multiple users etc					

this offering in a network realizing a system of collaborative systems in the various engineering departments [41].

Technology Model (Physical)—As built: The technology model of the platform is implemented with a redundant infrastructure taking stringent security aspects into account regarding access, data management and processing [42]. On the user perspective, it provides a number of user frontends to offered services as well as machine-to-machine communication for an automatic processing of collaboration workflows [43]. The user interface is thereby described using graphical mock-ups which are able to visualize results at any stage. Model-based Physical Data models with diagrams of the technology architecture; control structures, definitions and descriptions are realized using UML techniques [44, 45].

Detailed Representations: The use of Node templates, enable each customer of OpenDESC.Com to be set up easily as a node instance with its own specific characteristics including one for Digital Twin [46]. Each node is logically separated from other customers in order to fulfill stringent security requirements [47]. The operation of a node is conducted over an administration and operation interface, which allows its overall management.

Functioning Enterprise: Each authorized user can create communication relationships, define rules for exchange, timing, quantity, desired data quality, level of detail and protection of intellectual property [32], add required metadata, distribute data sets to multiple users etc.

10.7 Adoption of Extended Services

An efficient simulation set-up requires the fulfilment of some pre-requisites. One of them is high-quality input data. This helps to create simplified access to simulation models and their use [46]. Also, with this the simulation process can be optimized, it shows the potential for savings, and it enables productivity as well as quality increases [48]. The content of the extended services for generation of Digital Twin of manufacturing follows here.

10.7.1 Working Procedure

Based on the existing production system, the process for generation of the Digital Twin consists of four basic steps:

1. Scan of the production system.
2. Object recognition by using Machine Learning (ML).
3. Automatic simulation modeling.
4. Analyzes with the simulation model.

10.7.1.1 Scan

During the scanning process (as described in Chap. 4 in more detail), it is possible that the user can observe the progress via a sparse point cloud. This sparse point cloud is displayed over the filmed image. The user can see the progress by being able to track which area has already been recorded. The single images are streamed to a central system via WLAN or saved on the device. After filming, a 3D dense point cloud is generated. With this offline generation of the dense point cloud, it is possible to perform further post-processing (camera-pose-correction, bundle-adjustment) over the entire 3D point cloud and thus increase the accuracy. The accuracy of the 3D points is about one or two centimetres over a clear length of ten meters and is therefore sufficient for the specifications of the simulation model.

It also possible to film without a central system, in this case the records are stored on a camera device. If the record is finished, the user uploads the video to the cloud. The cloud system starts to generate the sparse point cloud. At this point, the user is able to check the sparse point cloud. If everything is acceptable, the dense point cloud generation is proceeded. The generation of the dense point cloud can take some hours or days of calculation. In summary, every scan has the following steps:

- Record of the production hall.
- Generating a sparse point cloud.
- Post-processing (image, camera-pose-correction, bundle-adjustment).
- Generating the dense point cloud.

10.7.1.2 Object Recognition by Using Machine Learning

With object recognition based on the point cloud (as described in Chap. 8), a parametric CAD model is selected from a pre-defined library with all relevant objects for each recognized object and inserted in the respective position in the hall. Unrecognized objects are replaced by enveloping cuboids (bounding boxes), which reduces reworking to replacing the bounding box with a real CAD model.

The step of object recognition generates the following information:

- Recognized objects (or also unrecognized objects, which are each replaced by a bounding box).
- Position of the objects in the room, hall boundaries such as walls and windows.
- Structure of the hall in nine categories.

With this information, an assembly is automatically generated in target CAD system (e.g. SolidWorks), which represents the complete hall with all objects as individual parts. This intermediate result can be used as input for simulation but also for other processes (e.g. change design of the production system).

10.7.1.3 Automatic Simulation Modeling

The simulation model is built automatically with the information from the object recognition. The information transferred is, on the one hand, the identified objects, including further information on these objects. The spatial positions of the objects are also transferred to the simulation software. On the other hand, a CAD model is transmitted to the simulation software so that the simulation model receives a graphical interface that is very close to a CAD model.

This procedure separates the function, i.e. the objects that determine the actual sequence of the simulation, and the visualisation. The advantage of this procedure is that the function can be transferred to the standard modules of the simulation software. Accordingly, the function does not have to be reprogrammed, but the prefabricated functionalities of the simulation software can be used. In addition, the integration of the graphical user interface offers the advantage that an accurate and detailed simulation model is created. This aspect is important for traceability and acceptance by the customer.

All steps are highly automated for standard processes, which means that simple production processes are included in the scope of the current programming. If, in contrast, there are company-specific peculiarities, such as special linking logics between the single processing stations or priority rules for production control, these must be reprogrammed as part of a service. A detailed description of this process can be found in Chap. 9 of this book.

10.7.1.4 Analyzes with the Simulation Model

Three types of analysis are distinguished. An initial consultation is necessary in which a specific analysis is discussed together. The three standard analyses are:

– **Bottleneck analysis**: Here, a list of (future) load of the workstations, incl. identification of free capacities and potential estimation in case of a possible increase of the bottleneck capacity, is determined. For this analysis, the customer must define the factory areas and the objects under consideration, accompany the scanning procedure (scheduling, presence at the scanning process) and provide operating data. The remaining steps are covered by the service. The scope of the service depends on the size of the factory areas and number of workstations, whether scanning can be done during ongoing operations and what level of detail is desired for the bottleneck analysis. Optionally, setup processes and material supply can be included in the analysis.

  Figure 10.5 shows a scheme for the bottleneck analysis. The load factor is determined for all stations. In this fictitious case, the high capacity load at Station I leads to a low value-added share at the subsequent stations. Accordingly, a fictitious result in this case would be that the increase in capacity at Station I leads to a productivity gain, so that the Digital Twin can be used to check whether a further parallel machine for Station I would be beneficial. Similarly, the proportion of

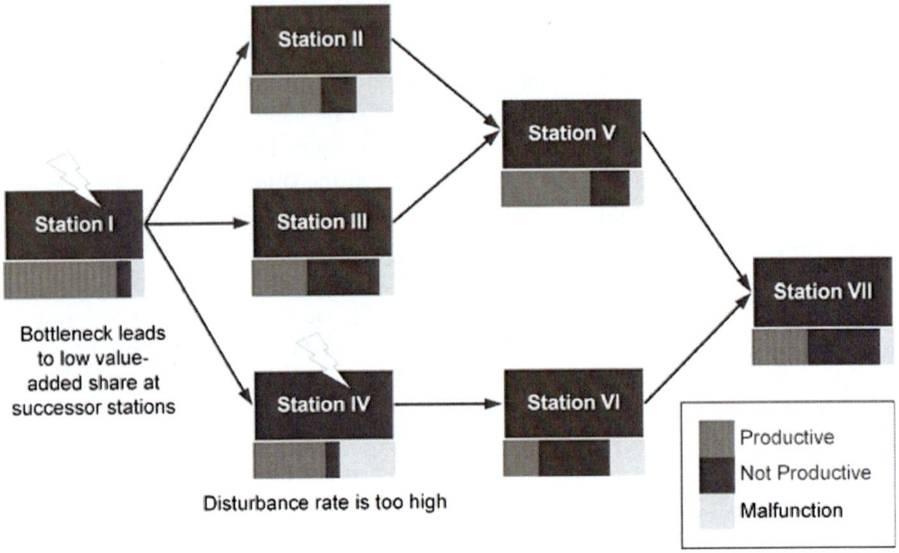

Fig. 10.5 Scheme for bottleneck analysis

malfunctions at Station IV is too high, so that Station VI is often poorly loaded. Here, it would be necessary to check whether the proportion of malfunctions can be increased through maintenance activities. This example clearly shows how a bottleneck analysis with the Digital Twin can be a benefit for a company.

- **Material flow analysis**: With this analysis, a material flow and value stream analysis, incl. Sankey diagram, stock level, processing times and identification of optimization potentials, is carried out. Here, too, the customer must select factory areas and the items under consideration, accompany the scanning process and provide the required operating data. The remaining steps and also the analysis are carried out in the service. The scope of the service again depends on the size of the factory areas and the number of workstations, whether scanning can be performed during ongoing operations and whether setup processes and other processes, such as material supply, should be considered.
  Figure 10.6 again shows a diagram illustrating an exemplary case of the analysis. Here a Sankey diagram is used to show the flow of products through the factory. Thereby, more products flow over a path when the thickness of the arrow increases. In addition, the stocks and the connections between the stations are considered and displayed here. In the lower area, the throughput time is depicted so that analyses can be carried out. Accordingly, it is possible to optimize the stocks, synchronize the processes and plan the material transport.

- **Investment planning**: Investment planning offers the customer the examination of any investment alternatives and a selection of the best option. Here, the customer is required to select the possible investment decisions (e.g. different machines) and to specify target values. For this, corresponding data such as the purchase costs and

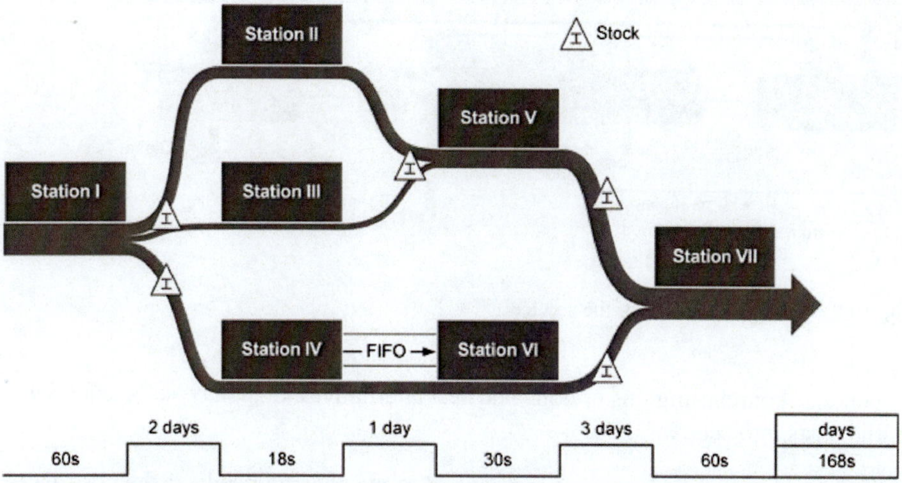

Fig. 10.6 Scheme of a material flow analysis

the expected new output are required. Likewise, the scan must be accompanied again and operating data must be provided. At this point, the scope of the service depends on the size of the factory areas and the number of investment decisions, as this is where the scenarios to be simulated arise. Likewise, the level of detail must be determined and the question arises as to whether scanning can take place during operation.

Figure 10.7 again shows a schematic of the analysis, this time for investment planning. Based on the specific production system and the selected alternatives, different scenarios are simulated. In the left part of the figure, three fictitious options are shown. For each option, the Digital Twin is used to calculate the

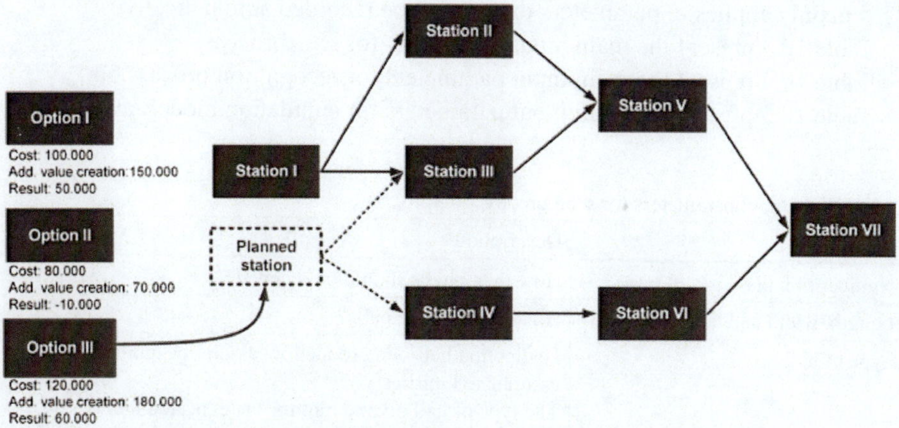

Fig. 10.7 Scheme of investment planning

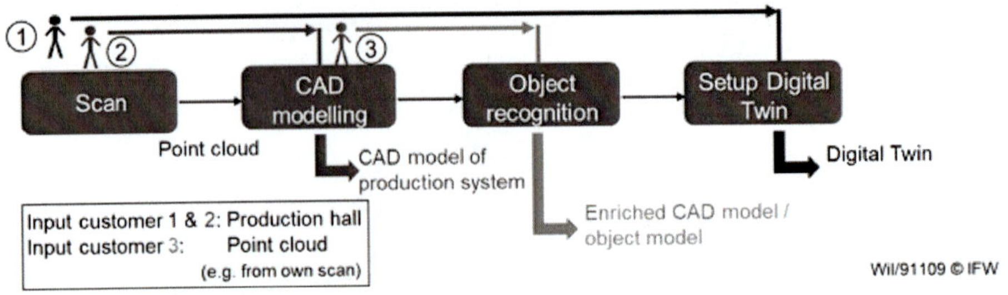

Fig. 10.8 Optional extension of the service

benefit of purchasing this option. The best alternative can then be selected, taking the costs into account.

In general, other analyses that correspond to the specific needs of the customers are also conceivable and can be implemented with the Digital Twin. For example, it is possible to purchase either the CAD model or the Digital Twin in the form of a simulation model. This means that customers can use the models themselves, provided they have the necessary knowledge in the field of IT and simulation tools. Figure 10.8 shows the different extensions of the service, related to the supply of whole models. Here, the client can receive different models according to the specific input. Likewise, as mentioned above, extensions are possible through arbitrary analyses.

10.7.2 *Input and Outcome*

Comprehensive planning of this service is necessary due to the diversity of the production environment. The relevant questions must be clarified before the scan. This mainly applies to parameters that cannot be recorded automatically.

Table 10.3 present the main input parameters for scan process.

Table 10.4 present the main input parameters for recognition process.

Table 10.5 present the main input parameters for simulation modeling.

Table 10.3 Input parameters for scan process

Value	Description
Number of halls	How many halls should be scanned?
Length, width and height per hall	How large are the halls?
Type of hall	Distinction between production, assembly, quality assurance, logistics The type of hall gives a hint for the expected number of objects in the hall
Shift operation	By specifying the shift operation, customer indicates when the scanning can be conducted

Table 10.4 Input parameters for recognition process

Value	Description
Plans	2D and 3D plans Mark on 2D plans which spatial parts are relevant (if not the whole hall) Mark areas with poor accessibility for the scanner (every object should be visible without the occlusion) Mark the position of the glass surfaces (due to shadowing) Buffer positions Manual working places Connected objects Transport routes
Inventory list	Number (approximately) of objects per group that are relevant for digitization • Machines • Conveyors • Charge carrier • Mobile parts • Furniture • Building parts • Supply lines • Which objects lie in the height larger than two meters
Markings on the objects	Object labels available (barcode, QR)?

Table 10.5 Input parameters for simulation modeling

Value	Description
Product data	Information on the products that are being manufactured, e.g. intermediate products, batch sizes, etc
Production program	Information on an exemplary or a previous production program, from which it emerges which products, at what intervals and in which batch sizes are being manufactured, so that the simulation process can be designed with it and scenarios can be generated
Work plan	In order to be able to map the process in production correctly, the work plans in a simplified form are needed, which shows which routes the products take through production. Any work instructions or internals are not needed, just the typical and alternative work stations for the products
Further data depends on results of analysis	Depending on which details are to be included in the simulation, additional data such as investment alternatives, information on material transport, buffers or setup processes are needed. Generally, however, this data is optional

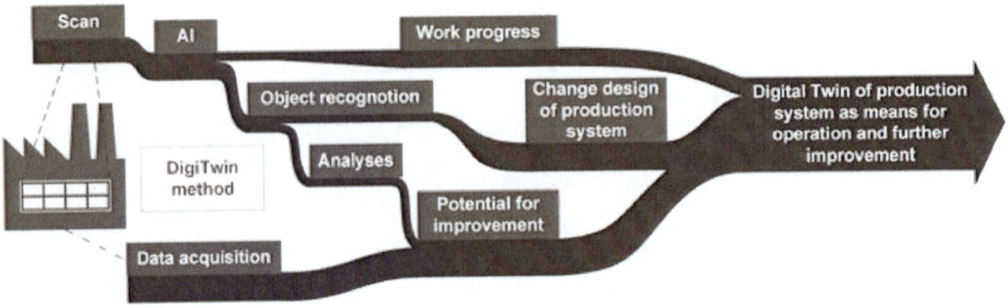

Fig. 10.9 Outcome of the DigiTwin method

The outcome of this service lies in Digital Twin itself as well as in intermediate results: results of various simulations, the CAD model of the production system and the hall, the adjusted point cloud for sake of work progress control. The outcome is shown in Fig. 10.9.

10.8 Adoption of Digital Twin Offering

Production systems change over time, for example when new machines are purchased or the existing layout is modified. This means that the Digital Twin created is constantly losing accuracy [49]. In turn, problems arise for the necessary updating process:

(1) Update costs are difficult to estimate.
(2) Lack of expertise in the use of simulation and IT tools.
(3) Unforeseeable operating expenses during use.

These obstacles result in the need for an efficient updating process for an existing Digital Twin of the production system in order to avoid the need for such updating processes in the future.

To overcome obstacle 1: The costs for an update process arise from high time efforts and manual processes that are difficult to calculate. Therefore, a simple procedure is needed in which the user company itself is enabled to initiate the update process by carrying out the scan independently. For this purpose, an application can be used to scan the areas to be updated in the Digital Twin using the camera of a commercially available smartphone [50]. The application guides the user through the scanning process so that it can be carried out quickly and easily.

To overcome obstacle 2: Even small changes to the Digital Twin require expertise in operating with the IT application [51]. The present research project relieves the user company of this necessity. The scan taken with the application is further processed in a highly automated way and the Digital Twin is updated with it. For this purpose, the data transfer with OpenDESC.Com (see Sect. 10.5) is used. The developed object recognition identifies the objects present in the scan. This means that the application does not require any further IT knowledge on the part of the user company.

To overcome obstacle 3: In the course of operation, further updating needs also arise, e.g. due to new products to be manufactured. However, the resulting processes cannot be captured via a scan, as they are not explicitly visible. This applies in particular to the production process by machine tools. Nevertheless, the specific behaviour of the machines can also be mapped by a Digital Twin of the machine, which makes it possible to predict processing times, malfunction profiles and quality parameters [52]. This anticipates selective updating processes for the behaviour of the machines. A scalable Digital Twin of production is created, in which Digital Twins of machines are embedded. In the sense of the automation pyramid, a vertical integration between the different planning levels of production is created, making the production planning and control of industrial companies more reliable, efficient and comprehensive [53]. This process is planned as an extension for the future and serves as an outlook at this point.

The use of a Digital Twin of manufacturing represents a tool with a very high market potential, as productivity can be increased for manufacturing companies and competitiveness can be secured in the long term. The Digital Twin provides decision support for strategic decisions and is simultaneously used in operational planning and control processes. A continuous updating of the Digital Twin is necessary to enable its ongoing operation. A one-time creation of the Digital Twin without a subsequent update option leads to companies shying away from the initial investment because they are aware of the constant changes in production. As a result, the digital twin quickly loses its value in their eyes. The service offered here enables the fast and inexpensive use and ongoing updating of a Digital Twin without additional IT expertise at the SME by simply scanning the area to be updated with a smartphone.

10.9 Discussion

Digital business ecosystems emerge as new means for value creation in networks where the self-organisation is controlled by digital infrastructure. Although the opportunities for profit and growth are obvious, the realization is highly challenging due to the disruptive changes in business model and organisation. However, the coping patterns must be identified for these key challenges to profit from newly evolving opportunities. The challenges lie on a technological, organisational and individual level. This yields an understanding of organisations as open but distinguishable systems with different interdependent entities providing complementary contributions. Complementarity between members is not a starting condition but needs to be established in a mutual development process. as firms have to handle the interconnectivity, the interoperability and the conflict of interests [26].

The multiplicity of actors in an ecosystem generates new trade-offs that direct the strategy. These trade-offs belong the cascade consisting of focal firm, its direct partners (i.e., traditional bilateral bargaining power), and their own partners to whom the focal firm may have no direct relationships. This yields deliberation, beyond the question of value creation and capture, to the question of value distribution across the broader ecosystem [27]. The idea of expanding the service portfolio is inherent in the concept of an ecosystem. It is a strategic lever in ecosystem structure which imposes the ecosystem dynamics, for example, when innovation requires a change in the configuration of ecosystem.

In context of the innovation design, designing networks tend to become design research networks which can lie in an ecosystem like OpenDESC.Com with the various ways of putting designing capacity into action, of "designing" and "being designers." By crossing the polarities between problem solving and sense making, and between expert and diffuse design, a map of the field of design modes can be obtained [54]. The map is built on two dimensions: the "actors and competence" axis from diffuse design to expert design, and the "motivations and expectations" axis from sense making to problem solving. By crossing them, each of four quadrants proposes a characteristic design mode and its recent evolutions (Fig. 10.10).

Grassroots organization is located in the diffuse design/problem solving quadrant and designs initiatives that aim to deal with local problems such as lack of green space in a neighborhood, difficulty of access to organic food, alternative mobility. Cultural activists occupy the diffuse design/sense making quadrant and are interested in cultural activities, who set up venues to promote their areas of interest and to create occasions for exhibiting, presenting and exchanging experiences, and debating. Design and communication agency fills the expert design/sense making quadrant by experts who use their specific knowledge and tools to conceive and develop original products, services, and communicative artifacts. This design modality includes most of the traditional design and communication agencies, many of which carry out their work focusing attention on consolidated products and services (from furnishings to hotels, from fashion to shops) [54].

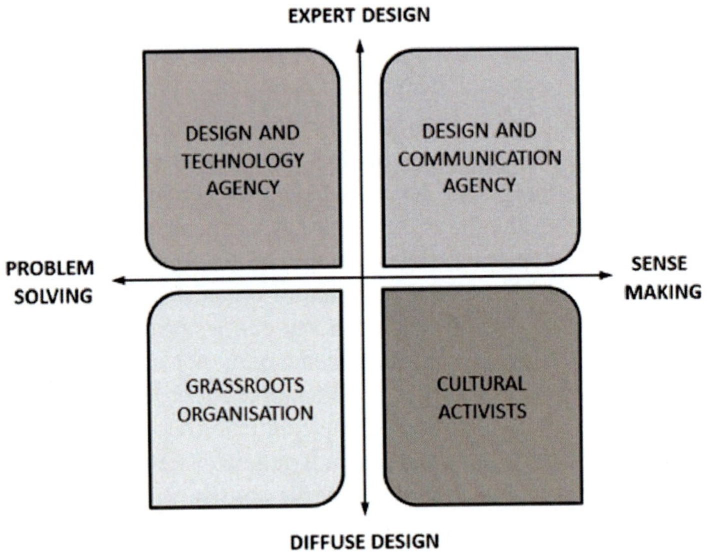

Fig. 10.10 Mapping design agency [54]

In contrast to previous three quadrants, the design and technology agency occupies the expert design / problem solving quadrant and represents the experts with a highly technical background, aiming at solving complex problems by bridging technical and social issues. This design mode is normally performed by design agencies based on transdisciplinary teams [55]. It addresses a variety of clients ranging from businesses to public bodies and citizen associations. Over the past decade it has been strongly influenced on one side by the user-centered design approach and methodology and by co-designing on the other [54]. That remains the place where the problem solving and the expert design meet each other what is in focus of OpenDESC.Com for decades.

Ecosystems as arrangements of interdependent value creation will only grow in prevalence and in importance in the coming years. In the world of practice, the embrace of the notion of ecosystems has been enthusiastic and, at the same time, chaotic [26]. In digital business ecosystems, the loosely coupled layers of digital objects enable firms to collaborate on a rather technical layer, for example, digital service, and compete on another layer, for example, customer-oriented solutions [27]. Critical to the success is a clear set of definitions, concepts, relationships, and boundaries. On an individual level, the role and cognition of managers and the competences of participating stakeholders to cope with the opportunities, heterogeneity, dynamics and complexity of product-service-systems are crucial for a successful co-creation [27].

10.10 Conclusions and Outlook

The aim of this chapter is to present how an innovative solution for generation of Digital Twin in the built environment can be commercialized as an extended service in an existing ecosystem by addressing new customer groups too. The environment in which this occurs is the manufacturing industry, with its deep supply chain, where innovations are stimulated at various levels [56]. Since many members of such a supply chain try to expand their business via a product or service platform, the particularly successful companies develop their own ecosystems that are particularly suitable for generating and disseminating further innovative concepts and solutions [57]. One of them is OpenDESC.Com, which has gained global acceptance over the past 23 years [29]. This enabled the Digital Twin to be added with less effort.

After the successful completion of the DigiTwin research project, the participating institutes and companies decided to work together on the industrialization of the project results and to design a joint solution for SMEs. As described in Chaps. 5–9, this solution covers the entire process chain from the creation of the Digital Twin to factory planning and simulation of the production processes and consists of four components: scanning, object recognition, CAD modeling and process simulation [58]. As a result, the division of labor from the project was retained. Furthermore, the presented solution was built using many existing components in this technological platform.

In addition, the partners hold introductory workshops to make the benefits of the process simulation clear to customers. The cost of setting up a Digital Twin can be clearly calculated through fixed price offers based on volume of the hall and the quantity of objects of interest with expected results and can be minimized through shared services when recording multiple production sites. Additional requirements such as iterative simulations and consequences of design changes to optimize the manufacturing processes are covered by the wide range of services offered by the partners. An additional business field in plant construction was also identified.

Material flow simulations for factory, investment, capacity and inventory planning, bottleneck analyzes and in-house material transport help improve company processes [59]. Up until now, the generation of the corresponding simulation models was very complex, which made it difficult for small and medium-sized companies to use them. However, digitization offers new possibilities to simulate and optimize the real situation in production with the help of a Digital Twin. How such simulation models can be created largely automatically from scan data has been researched in depth and developed into a comprehensive range of services. It enables companies to gain important insights to optimize their factory processes. Without having to build up your own IT expertise, you will be able to make well-founded operational decisions. This enables you to achieve robust production with little control effort, increase transparency and increase your productivity.

References

1. Stark J (2019) Product lifecycle management: paradigm for 21st century product realisation, vol 1, 4th edn. Springer International Publishing, Cham. https://doi.org/10.1007/978-3-030-28864-8
2. Biahmou A, Majić T, Stjepandić J, Wognum N (2019) Platform-based OEM-supplier collaboration ecosystem development. Adv Transdiscipl Eng 10:436–445
3. Katzenbach A (2015) Automotive. In: Stjepandić J et al (eds) Concurrent engineering in the 21st century: foundations, developments and challenges. Springer International Publishing, Cham, pp 607–638
4. Sillanpää I, Shahzad K, Sillanpää E (2015) Supplier development and buyer-supplier relationship strategies—a literature review. Int J Procure Manag 8(1/2):227–250
5. Taherdoost H, Brard A (2019) Analyzing the process of supplier selection criteria and methods. Procedia Manuf 32:1024–1034
6. Zhang R, Wang K (2019) A multi-echelon global supply chain network design based on transfer-pricing strategy. J Ind Integr Manag 4(1):1850020. https://doi.org/10.1142/S2424862218500203
7. Rupčić N, Majić T, Stjepandić J (2020) Emergence of business ecosystems by transformation of platforms through the process of organizational learning. J Ind Integr Manag 5(2):181–203
8. Bortolaz Guedes G, Zattar Paganin LB, Borsato M (2018) Bibliometric and systemic analysis on material flow mapping and industrial ecosystems. J Ind Integr Manag 3(4):1850001. https://doi.org/10.1142/S242486221850001X
9. Morelli N, de Götzen A, Simeone L (2021) Service design capabilities. Springer Nature Switzerland AG, Cham
10. Kimbell L (2011) Designing for service as one way of designing services. Int J Des 5(2):41–52

11. Bruglieri M (2018) Multidisciplinary design of sharing services. Springer International Publishing, Cham
12. OECD (2015) Frascati manual 2015: guidelines for collecting and reporting data on research and experimental development. The Measurement of Scientific, Technological and Innovation Activities, OEMD Publishing, Paris. https://doi.org/10.1787/9789264239012-en
13. Peng X, Cai L, Lu S, Cai Y, Gao Y (2016) Antecedent and dimension of symbiotic relationship in the hub-based entrepreneurial ecosystem: case study of Alibaba. J Ind Integr Manag 1(4):1650011. https://doi.org/10.1142/S2424862216500111
14. Bondar S, Hsu JC, Stjepandić J (2015) Network-centric operations during transition in global enterprise. Int J Agil Syst Manag 8(3–4):355–373
15. Zhang R, Wang K (2019) Service supply chain research: a conceptual model based on business processes. J Ind Integr Manag 04(03):1950007. https://doi.org/10.1142/S2424862219500076
16. Cook M, Mo JP (2019) Architectural modeling for managing risks in forming an alliance. J Ind Integr Manag 4(3):1–17
17. Bondar S, Stjepandić J (2018) Engineering collaboration in product development of modular products. Adv Transdiscipl Eng 7:391–400. https://doi.org/10.3233/978-1-61499-898-3-391
18. Raudberget D, Elgh F, Stolt R, Johansson J, Lennartsson M (2019) Developing agile platform assets—exploring ways to reach beyond modularisation at five product development companies. Int J Agil Syst Manag 12(4):311–331. https://doi.org/10.1504/IJASM.2019.104588
19. Bork D, Buchmann RA, Karagiannis D, Lee M, Miron E (2019) An Open platform for modelling method conceptualization: the OMiLAB digital ecosystem. communications of the association for information systems, vol 44, pp 673–697. https://doi.org/10.17705/1CAIS.04432.
20. Moynihan P, Dai W (2011) Agile supply chain management: a services system approach. Int J Agile Syst Manag 4(3):280–300
21. Mo JPT, Beckett RC (2019) Architectural modelling of transdisciplinary system with inherent social perspectives. J Ind Integr Manag: Innov Entrep 4(4):1950012
22. Harland PE, Uddin Z, Laudien S (2020) Product platforms as a lever of competitive advantage on a company-wide level: a resource management perspective. RMS 14:137–158. https://doi.org/10.1007/s11846-018-0289-9
23. Lee YY, Meurk CS, Harris MG, Diminic S, Scheurer RW, Whiteford HA (2014) Developing a service platform definition to promote evidence-based planning and funding of the mental health service system. Int J Environ Res Public Health 11(12):12261–12282. https://doi.org/10.3390/ijerph111212261
24. Zhu F, Furr N (2016) Products to platforms: making the leap. Four steps to successful transformation, Harvard busines review. Accessed 10 Jan 2021. https://hbr.org/2016/04/products-to-platforms-making-the-leap
25. Hein A, Schreieck M, Riasanow T, Setzke DS, Wiesche M, Böhm M, Krcmar H (2020) Digital platform ecosystems. Electron Mark 30:87–98. https://doi.org/10.1007/s12525-019-00377-4
26. Lenkenhoff K, Wilkens U, Zheng M, Süße T, Kuhlenkötter B, Ming X (2018) Key challenges of digital business ecosystem development and how to cope with them. Procedia CIRP 73:167–172. https://doi.org/10.1016/j.procir.2018.04.082
27. Adner R (2017) Ecosystem as structure: an actionable construct for strategy. J Manag 43(1):39–58. https://doi.org/10.1177/0149206316678451
28. Rong K, Lin Y, Li B, Burström T, Butel L, Yu J (2018) Business ecosystem research agenda: more dynamic, more embedded, and more internationalized. Asian Bus Manag 17(3):167–182
29. Bondar S, Potjewijd L, Stjepandić J (2012) Globalized OEM and tier-1 processes at SKF. In: Stjepandic J et al (eds) Advanced concurrent engineering. Springer, London, pp 789–800
30. Bondar S, Bouwhuis H, Stjepandić J (2015) Engineering collaboration in mechatronic product development. Adv Transdiscipl Eng 2:565–574. https://doi.org/10.3233/978-1-61499-544-9-565
31. Stjepandić J, Liese H, Trappey AC (2015) Intellectual Property Protection. In: Stjepandić J et al (eds) Concurrent engineering in the 21st century: foundations. Developments and Challenges, Springer International Publishing, Cham, pp 521–552

32. Liese H, Stjepandic J, Rulhoff S (2010) Securing product know-how by embedding IP-protection into the organisation. In: 2010 IEEE international technology management conference, ICE 2010, 7477025
33. Biahmou A, Stjepandic J (2016) Towards agile enterprise rights management in engineering collaboration. Int J Agil Syst Manag 9:302–325
34. Jiang H, Gao S, Zhao S, Chen H (2020) Competition of technology standards in industry 4.0: an innovation ecosystem perspective. Syst Res Behav Sci 37:772–783. https://doi.org/10.1002/sres.2718
35. Bondar S, Shammaa A, Stjepandić J, Tashiro K (2015) Advances in parameterized CAD feature translation. Adv Transdiscipl Eng 2:615–624. https://doi.org/10.3233/978-1-61499-544-9-615
36. Bondar S, Ruppert C, Stjepandić J (2014) Ensuring data quality beyond change management in virtual enterprise. Int J Agil Syst Manag 7(3/4):304–323
37. McLay A, Everett R, Keith-Storey A (2016) Devising a quality management system for new service-support contracts in industry. Int J Agil Syst Manag 9(4):340–362
38. Bondar S, Hsu JC, Pfouga A, Stjepandić J (2017) Agile digital transformation of system-of-systems architecture models using Zachman framework. J Ind Inf Integr 7:33–43
39. Viriyasitavat W, Bi Z (2020) Service selection and workflow composition in modern business processes. J Ind Inf Integr 17. https://doi.org/10.1016/j.jii.2020.100126
40. Beckett RC, Vachhrajani H (2017) Transdisciplinary innovation: connecting ideas from professional and user networks. J Ind Integr Manag 2(4):1750016. https://doi.org/10.1142/S2424862217500166
41. Mathiasen JB, Mathiasen RM (2017) Practicing transdisciplinary engineering in a global development context: the transferring, translating and transforming approaches. J Ind Integr Manag 2(4):1750017. https://doi.org/10.1142/S2424862217500178
42. Chen Y, Dong F, Chen H (2016) Business process and information security: a cross-listing perspective. J Ind Integr Manag 1(2):1650009. https://doi.org/10.1142/S2424862216500093
43. Fraga AL, Vegetti M, Leone HP (2020) Ontology-based solutions for interoperability among product lifecycle management systems: a systematic literature review. J Ind Inf Integr 20, 100176. https://doi.org/10.1016/j.jii.2020.100176
44. Gorkhali A, Xu L (2017) Enterprise architecture: a literature review. J Ind Integr Manag 2(2):1750009. https://doi.org/10.1142/S2424862217500099
45. Gorkhali A, Xu L (2019) Enterprise architecture, enterprise information systems and enterprise integration: a review based on systems theory perspective. J Ind Integr Manag 4(2):1950001. https://doi.org/10.1142/S2424862219500015
46. Kuhn O, Liese H, Stjepandić J (2011) Methodology for knowledge-based engineering template update. In: Cavallucci D, Guio R, Cascini G (eds) Building innovation pipelines through computer-aided innovation. Springer, Berlin Heidelberg, pp 178–191
47. Biahmou A, Stjepandić J (2016) Towards agile enterprise rights management in engineering collaboration. Int J Agil Syst Manag 9(4):302–325
48. Symeonidou N, Bruneel J, Autio E (2017) Commercialization strategy and internationalization outcomes in technology-based new ventures. J Bus Ventur 32:302–317. https://doi.org/10.1016/j.jbusvent.2017.02.004
49. Denkena B, Doreth K, Köller M, Wilmsmeier S, Winter F (2020) Simulationsbasierte kombinierte Instandhaltungs- und Produktionsplanung. In: Meyer G, Pöge C, Spieckermann S, Wenzel S (eds) Ablaufsimulation in der Automobilindustrie. Springer Vieweg, Berlin
50. Kamble K, Kulkarni H, Patil J, Sukhatankar S (2018) Object recognition through smartphone using deep learning techniques. In: 2nd international conference on soft computing systems, ICSCS 2018, communications in computer and information science, vol 837, pp 242–249
51. Al-Zubaidi SQD, Fantoni G, Failli F (2021) Analysis of drivers for solving facility layout problems: literature review. J Ind Inf Integr 21, 100187. https://doi.org/10.1016/j.jii.2020.100187
52. Liu J, Zhou H, Tian G, Liu X, Jing X (2019) Digital twin-based process reuse and evaluation approach for smart process planning. Int J Adv Manuf Technol 100:1619–1634. https://doi.org/10.1007/s00170-018-2748-5

53. Meudt T, Pohl M, Metternich J (2017) Die Automatisierungspyramide-Ein Literatur-überblick
54. Manzini E (2015) Design, when everybody designs an introduction to design for social innovation. The MIT Press, Cambridge
55. Wognum N, Bil C, Elgh F, Peruzzini M, Stjepandić J, Verhagen W (2018) Transdisciplinary engineering research challenges. Adv Transdiscipl Eng 7:753–762
56. Sprong N, Driessen PH, Hillebrand B, Molner S (2021) Market innovation: a literature review and new research directions. J Bus Res 123:450–462. https://doi.org/10.1016/j.jbusres.2020.09.057
57. Oster C, Wade J (2013) Ecosystem requirements for composability and reuse: an investigation into ecosystem factors that support adoption of composable practices for engineering design. Syst Eng 16(4):439–452. https://doi.org/10.1002/sys.21256
58. Sommer M, Stjepandić J, Stobrawa S, von Soden M (2020) Automated Generation of a digital twin in manufacturing for a built environment using scan and object detection. J Ind Inf Integr XX, in press
59. Sommer M, Stjepandic J, Stobrawa S, Von Soden M (2020) Improvement of factory planning by automated generation of a digital twin. Adv Transdiscipl Eng 12:453–462. https://doi.org/10.3233/ATDE200105

Chapter 11
Digital Twin: Conclusion and Future Perspectives

Josip Stjepandić, Markus Sommer, and Sebastian Stobrawa

11.1 Introduction

The foundations, methods, and recent developments for generation of the Digital Twin and its applications in manufacturing have been presented and discussed at length in the preceding Chapters of this book. As a part of the respective Chapters in the discussion sections, authors have presented and discussed specific research and practical challenges of a specific process chain or step in isolation focused on a dedicated project goal. The aim of this final Chapter is twofold: at first, to draw the way for the further development of the DigiTwin solution as presented in this book, and, at second, to integrate and present current trends and challenges with respect to Digital Twin and its associated research (sub)domains, in particular Artificial Intelligence, in a comprehensive overview. To perform this integration in a structured, methodical manner, the socio-technical dimensions of Digital Twin are also explored. Recent research directions in non-technical areas (medicine, social science, economy) show further widespread of Digital Twin as a holistic, domain-independent approach. With this procedure, authors follow the structure for the final chapter already developed in prior research on Concurrent Engineering [1] and Systems Engineering [2].

J. Stjepandić (✉)
PROSTEP AG, Darmstadt, Germany
e-mail: josip.stjepandic@prostep.com

M. Sommer
isb innovative software businesses GmbH, Otto-Lilienthal-Strasse 2, 88046 Friedrichshafen, Germany
e-mail: sommer@isb-fn.de

S. Stobrawa
Institut Für Fertigungstechnik Und Werkzeugmaschinen, Leibniz Universität Hannover, An der Universität 2, 30823 Garbsen, Germany
e-mail: stobrawa@ifw.uni-hannover.de

J. Stjepandić et al. (eds.), *DigiTwin: An Approach for Production Process Optimization in a Built Environment*, Springer Series in Advanced Manufacturing,
https://doi.org/10.1007/978-3-030-77539-1_11

For presentation and discussion of current trends and challenges in Digital Twin, a stable taxonomy with dedicated dimensions which can be described with few characteristics is beneficial. This set of dimensions helps to generalize the conclusions and to prevent the frequent entanglement of Digital Twin descriptions with specific industries and vast array of application domains. The used dimensions are based upon taxonomy of Digital Twin developed by Otto et al. [3] and are as follows (see Sect. 3.2):

- Data link
- Purpose
- Conceptual elements
- Model accuracy
- Interface
- Synchronization
- Data input
- Time of creation.

Although these eight dimensions are not exhaustive, these make up an assessment framework that facilitates processes to configure, implement and exploit a specific expression of Digital Twin according to a particular strategy. This procedure is focused more on the application rather on the architecture of a dedicated solution. Subsequently, benefits of specific Digital Twins can be identified easier. Thus, a taxonomy provides a valuable contribution to fill that research gap, as it helps to demarcate Digital Twin concepts from each other. The fulfilment of singular dimensions based on the recent review is highlighted in the Table 11.1 which is an extension of Table 3.1 [3].

Based on the comprehensive literature review, a particular, central type of a Digital Twin can be identified (highlighted in grey), whose characteristics, in sum, are named most frequently [3]. Therefore, most known implementations of Digital Twin discover the following characteristics (Table 11.1):

Table 11.1 Practice of Digital Twin, derived from taxonomy [3]

Dimension	Characteristics		
Data link	One-directional		Bi-directional
Purpose	Processing	Transfer	Repository
Conceptual elements	Physically independent		Physically bound
Accuracy	Identical		Partial
Interface	M2M		HMI
Synchronization	With		Without
Data input	Raw data		Processed data
Time of creation	Physical part first	Digital part first	Simultaneously

- A bi-directional data link is given if data flow happens simultaneously between digital and physical representation in both directions. Some authors see it as a mandatory part of a Digital Twin [4–6].
- Purpose of a DT is mostly the processing of data (monitoring, analysis, forecasting, or optimization) [7–9].
- Conceptual elements: although a Digital Twin is mostly directly bound to its physical counterpart in a one-to-one ratio, the occurrence of an independent relationship is significant. In such a case, a Digital Twin can be seen in combination with other physical systems or one system can possess multiple Digital Twins [10–12].
- Physical object is mostly described with an identical accuracy, although the partial (variable) accuracy makes sense for an efficient representation and fast processing [13–15].
- Depending on the purpose and working environment (e.g. fully automated production), the implemented Digital Twins posse either a machine-to-machine interface or a human–machine interface or both [16–18]. That is no characteristic of a substantial distinction.
- Although a synchronization between the Digital Twin and the physical counterpart is not always mandatory to build a Digital Twin, it is mostly supposed by research that a Digital Twin frequently obtains data updates during its lifecycle [19] or a synchronization occurs between the Digital Twin and the physical counterpart [20, 21].
- Data are the "fuel" of Digital Twin [22]. Therefore, Digital Twins receive input data from sensors or databases, either directly as pure, raw data from data acquisition devices or pre-processed (e.g., by analytic software) before use in the Digital Twin. Here, a substantial distinction only could be made in a further breakdown of data processing [23–25].
- Analysing the time of creation which describes the chronological order in which the respective parts of the Digital Twin come into existence, it becomes evident that the physical counterpart mostly is being developed first [26, 27]. That indicates that a Digital Manufacturing Twin and a Digital Instance Twin prevail in the practical exploitation of Digital Twin [28]. This can also be explained by the fact that the Digital Twin is subsequently generated for durable goods [29]. In this sense, Digital Twin can be understood as a means for digitalization of long-lasting built environment (see Chaps. 4 and 8) [30].

Besides of the above-explained eight dimensions, further views and aspects play an important role in the conception, the implementation and the exploitation of the Digital Twin [31]. Data volume and complexity caused by new concepts and associated domains such as Cyber-Physical Systems (CPS) and the Internet of Things (IoT) demand that a Digital Twin continuously works with actual data [32]. This yields methods and architectures for modelling and simulation with high level of mutual interactions [33]. These developments increasingly require an interdisciplinary or transdisciplinary approach to solve Digital Twin challenges, as the underlying constructs are themselves inter- or transdisciplinary (e.g., include new types of labour organisation or intercultural collaboration) [34]. The implementation is reinforced through different levels of aggregation, across system scope and time [35]. With respect to time, the shift of industry and academia towards an integrated perspective on products and services implies that a full lifecycle perspective on development is becoming prevalent [36]. The perspective of Digital Twin is mostly holistic: it remains active during the entire product lifecycle rather than constrained to a singular phase [37].

For a rapid emerging system like a Digital Twin, different levels of technological uncertainty require different levels of integration [38]. Integration difficulties become prominent, however, in the higher uncertainty type implementation projects. Digital Thread poses a solution for a continuous connection of all digital models over the entire product lifecycle phases [39]. It impacts recent trends regarding digitization such as consistency and traceability of data (see Chap. 7). In addition, there are consequential issues of configuration and risk management [40].

The outline of the chapter is as follows. In Sect. 11.2, the research and development directions of the DigiTwin solution presented in this book are drawn and discussed. In Sects. 11.3 and 11.4 self-X-mechanisms for experts and non-experts are presented. In Sect. 11.5 the trends of Digital Twin are discussed, following the eight dimensions given above. In a similar fashion, Sect. 11.6 highlights challenges in research and practice. Finally, a demarcation from related areas and conclusion on the presented work is given in Sect. 11.7.

11.2 Further Developments of DigiTwin Solution

This book shows how a Digital Twin of a production system can be efficiently created. The focus of the explained method is on the initial set-up, which means that a Digital Twin can be created once with the method, but no further adjustment of the Digital Twin is carried out after the method has been performed. The operation of the Digital Twin can take place from this point on, but production systems change over time. The following aspects are particularly relevant here, although other changes to the production system are of course also conceivable:

- Adjustment activities change the production sequences, which in turn changes processing times, for example. This is caused by general aims of production management to increase productivity through minor adjustments. However, negative productivity developments are also conceivable if, for example, higher quality requirements occur or machines become worn out.
- The production schedule may change over time. This input variable for the production system, meaning which product is to be produced at which time, is usually subject to constant alterations. However, major changes can also occur, for example due to new customers or modified sales strategies, so that the changes in the production schedule have an impact on the production system and its structure.

- New products or product variants cause adjustments to be made to the production processes or new ones to be developed. In the simplest case, the changes that arise for the Digital Twin may involve the insertion of new products and the specification of processing times, or in the most extensive case they may lead to a layout adjustment of the production system.
- Besides products, new machines may be added to the production system or old machines may be removed from the production system. This leads to new product movements through the production system, resulting in a comprehensive change.
- The layout of the production system can change by moving objects in the production hall. This results in changed product movements, which in particular changes the transport times.

The aforementioned changes mean that the resulting Digital Twin must be updated over time in order to continue to represent a valid representation of the real production system. The relevance of this aspect was presented by Denkena et al. [41]. However, updating the Digital Twin is not part of the method presented here in the book. Nevertheless, the aspect of updating should be discussed in this section in the form of an outlook, since the updating process was always considered during the development of the method and the method as such has requirements for updating. In order to illustrate the updating process, three general cases are explained here, which involve different levels of complexity or effort for updating and should therefore be considered separately from each other. In ascending order of complexity, these cases are:

1. *Parameter-based updating of the Digital Twin*: The changes in the real system lead to the requirement that the Digital Twin can remain in its current form, but that single parameters have to be updated.
2. *Structural partial update of the Digital Twin*: The changes in the real system require an adjustment of the Digital Twin, but the general structure can be preserved so that a new scan is not necessary or only isolated areas have to be re-recorded.
3. *Structural update of the Digital Twin*: The changes in the real production system are so extensive that a new recording is required and the Digital Twin has to be completely recreated.

11.2.1 Parameter-Based Updating of the Digital Twin

The first case mentioned is for example when, as mentioned above, the production schedule is changed or adjustments to the production system lead to altered processing times. Moreover, new products that do not lead to structural modifications of the production system can also be addressed with this case. It is therefore a matter of adjusting parameters of the Digital Twin, such as processing or set-up times.

For this case, update options are already provided in the method presented in this book. For example, with predefined user interfaces, new products can be inserted simply and without knowledge of the simulation software. Machining and set-up times can be added or updated in the Digital Twin via the work plans that can be stored here. The menu navigation therefore also allows existing products or the production schedule to be adapted. These processes and functionalities were presented in detail in Chap. 9 of this book.

A useful addition for this case is the connection of the Digital Twin to the existing IT systems, especially Enterprise Resource Planning (ERP) or Manufacturing Execution Systems (MES). However, a connection to machine or operating data is also reasonable at this point, as for example explained by Donhauser et al. [42]. With this interface connection to ERP or MES, information about products, production schedules, work plans, machining and set-up times, or similar can be transmitted directly to the Digital Twin. The updating of the Digital Twin is carried out automatically here, so that this process does not require any further handling and thus provides a practical extension to the method presented.

For the interface between ERP and MES to the Digital Twin, a data transfer is also provided in the method described in this book. Here, transfer tables were placed in the simulation software for each product and production schedule, which have a defined formatting. This formatting can be utilised for an ERP or MES export in order to transfer the data in a form so that it can be directly integrated into the Digital Twin. This procedure was also prototypically tested in a use case, but here the MES export had to be pre-processed before it was transferred to the simulation software in order to achieve the desired form of the data. For this purpose, a macro was developed in MS Excel, so that the data transfer can be highly automated. However, it would be desirable in the next step if the MES export were directly available in the defined formatting rule. This is a planned further step.

The simplest update process described here, which only involves parameter adjustment, is thus achievable in the next development steps. The basic functionalities are considered in the current method. However, it should be noted at this point that due to the large number of different MES and ERP, a generic solution can only be achieved with standardised transfer protocols.

11.2.2 Structural Partial Update of the Digital Twin

A much more far-reaching updating process occurs when, in addition to the parameters of the Digital Twin, the structure must also be adapted. This is always the case when objects, such as machines, are added, removed or displaced in the real production system. However, this case can also arise in the case of comprehensive changes to products or the production schedule, if this results in major modifications to the production system that cannot be mapped with a pure parameter adjustment of the Digital Twin.

A structural update can of course be carried out by people who have expertise in the simulation software. However, this manual process would be costly as well as fault-prone and the corresponding IT expertise must be available. In terms of the method described in this book, an efficient method would therefore lead to a better result, since again no IT expertise is needed, the effort is low, and the costs for the update are predictable.

For this reason, an expansion of the method is being aimed at for the next development steps, which will be presented below as an outlook. Accordingly, this presentation is a research concept at this point in time without any current findings. The following Fig. 11.1 provides an overview of the research approach.

The approach consists of three key aspects, shown in the Fig. 11.1, which are needed to proceed from a current Digital Twin (left hand side of the figure) to an updated Digital Twin (right hand side of the figure):

(1) It is intended that the change in the real system is recorded by the customer himself which is particularly beneficial if the customer is located far abroad. This means that if, for example, a machine is moved to a different location in production hall or a new machine is added to the production system, only this changed area is scanned. In this way, the overall effort is being reduced to a few objects of interests.
To be able to do this by a customer, a simple process with a simple easy-to-use device is needed. Here, an application for common smartphones is developed with which a production area can be recorded. Photogrammetry is used to generate the required point cloud. During the recording, the customer receives augmented reality feedback on whether the respective area has been correctly recorded by displaying objects in green on the mobile phone display if they have been captured. Finally, the recording can be sent directly to the offices that will further process the recording. For this purpose, the secure data transfer is used, which was presented in Chap. 7. The volume of data which is being generated and transferred here is significantly lower than for scan of the entire production system. If the initial result is not sufficient, this procedure can be repeated as often as necessary.

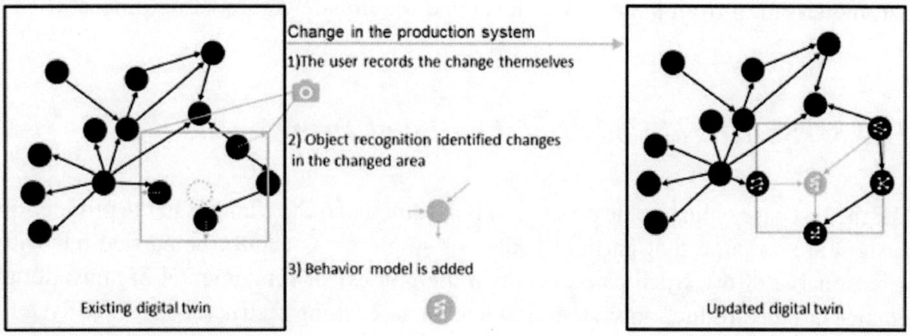

Fig. 11.1 Research approach for the structural update of the Digital Twin

(2) Accordingly, a simple application for a smartphone is created here that enables the client to take the scan of the changed areas himself. The service offered here is a lean process that can be carried out quickly without the need to visit the customer on site.
(3) The second aspect is the object recognition of the changed areas and the adaptation of the Digital Twin. Here, the method presented in this book must be adapted so that single areas from the Digital Twin can be segmented and modified. This poses a particular challenge because the edges between the existing and the modified areas have to be designed to fit each other. It must be resolved by an additional registration.
(4) In addition, however, this new updating method is quite similar to the method already explained for the initial creation of the Digital Twin. Here, too, a point cloud is first created from the scan (but here by the customer). Subsequently, object recognition identifies new objects and inserts them into the CAD model, which in a further step is transferred to the simulation software to adjust the Digital Twin.
(5) The last aspect involves an extension that mainly relates to the parameter-based updating of the Digital Twin. Likewise, this aspect represents an extension of the method presented in this book. The core idea is that behavioural models are stored for the single objects, especially machine tools. These behavioural models represent the functions and capabilities of the objects, so that changed conditions, such as new products, wear, malfunctions, etc., are directly represented by the models and do not require any further updating process for the Digital Twin. Accordingly, this aspect prevents the requirement for updating processes for the Digital Twin.

With the three aspects presented, there is therefore an efficient updating process that represents a sensible extension of the existing method. A scalable Digital Twin of manufacturing is created in short time with low effort, in which behavioural models of machines are embedded. In the sense of the automation pyramid, a vertical integration between the different planning levels of manufacturing is created, making the production planning and control of industrial companies more reliable, efficient and comprehensive [43]. The resulting Digital Twin becomes more accurate through parameterisation from the machine level and requires fewer updating processes.

11.2.3 Structural Update of the Digital Twin

For the last case of an update process, it is assumed that the changes in the production system are so large that partial updates of single areas cannot be carried out with a reasonable effort. Such cases occur in the context of fundamental organisational changes to the production system or when a far-reaching restructuring of the system is carried out.

The partial update process described in Sect. 11.2.2 is then too costly, as the additional effort required to assemble existing and updated parts of the Digital Twin becomes too great. Accordingly, in such a case, a renewed implementation of the method from this book would be recommended. This means that the complete factory floor is scanned again (see Chap. 4) and the Digital Twin is recreated with the upstream object recognition (see Chap. 8).

It should be emphasised at this point, however, that most processes can be shortened here, since an initial creation process already exists. Accordingly, company-specific data on products, for example, no longer needs to be fundamentally re-entered, but can be taken over from previous processes. Another useful extension is the integration of the machine models, as described in Sect. 11.2.2. This additional aspect then prevents the requirement for further updates.

By using the taxonomy as explained in Sect. 3.2, the classification of this new approach can be updated from status given in Sect. 3.6 as follows. With regard to data link, this Digital Twin becomes conditionally bidirectional (in case of update). Dimension purpose remains for data transfer and repository. Dimension accuracy remains partial (e.g., adjustable according to the process requirements). Dimension interface is still machine-to-machine. Synchronization between the physical and the digital part is not existent. Data input is fed with raw data. Time of creation becomes changeable. At first, it is pre-defined by the built environment: the physical counterpart first and then changed to digital part and vice versa.

11.3 Self-X Digital Twin on Backend

Advanced production systems need to interact with the physical and social world, as systems explicitly react to human behaviour and environmental conditions—and influence both through their actions. Thus, any two (sub-)systems coming into contact with the same part of the environment or group of users can influence each other in ways that can hardly be anticipated at design-time. Undesired effects may be the consequence of unintended, implicit interaction of elements that were not fully predictable or relevant beforehand. When designing Digital Twins, respective mechanisms have to be implemented, making the system autonomic as a combination of self-capabilities, such as self-configuration, self-healing, self-optimization, self-protection, self-awareness, etc [44]. This approach makes a system self-adaptable and self-decision-making support system for various activities [46].

The Self-X-mechanism of Digital Twins are suited to increase the value of production systems in operation [45]. Subsequently, the focus lies on the continuous improvements on different objectives. For example, energy or power optimization can be evaluated under different conditions. Other optimization scenarios can include:

- Reduced breakdown
- Reduced vibration

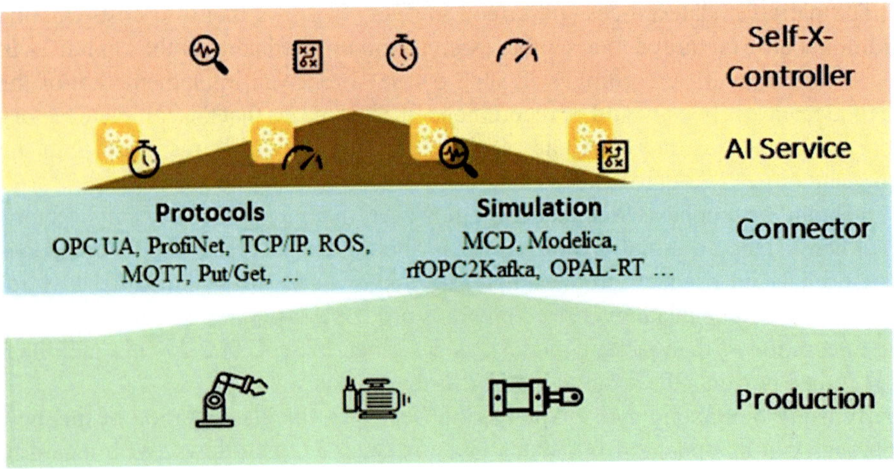

Fig. 11.2 Layers from production to Self-X-Controller

- Increased mechanical life
- Reduced fuel consumption.

The Digital Twin describes a production system including all necessary characteristics down to sub-systems of individual units. Therefore, the IT components of mechatronic systems comprise the control system in the physical machine and its entire Digital Twin. With the Self-X architecture another component is added: The Self-X-Controller being responsible for deciding on and initiating the Self-X activities. As the Self-X concept adds a lot of new functionality, a new architecture is needed for extending the Digital Twin concept with Self-X-Controllers (Fig. 11.2) The production layer communicates through connecting layer with AI services, sending real or simulated data to AI applications. While the AI layer evaluates individual processes, the additional Self-X-Controller layer tracks superordinate trends. If the difference between interpreted physical sensor readings and digital objectives becomes too high, the Self-X-Controller begins to intervene. The architecture must be scalable, distributed, and flexibly adaptable for different use cases in production [46].

For an efficient implementation of this new architecture, information (protocols of production processes, simulation results, machine states, etc.) must be transferred in real-time, making it accessible for the Artificial Intelligence (AI) services.

With this data, the AI service generates an AI model. If its accuracy is sufficient, the AI model is directly downloaded to the Edge Device at the machine and executed within the production environment. During production, AI models predict all values and the Self-X-Controller assesses the difference between the real and the predicted values. If the difference is too high, the Self-X-Controller sends the data to the backend, a supporting higher-level IT-infrastructure. If necessary, the AI model must be trained including the most recent data to increase accuracy. If the accuracy of the

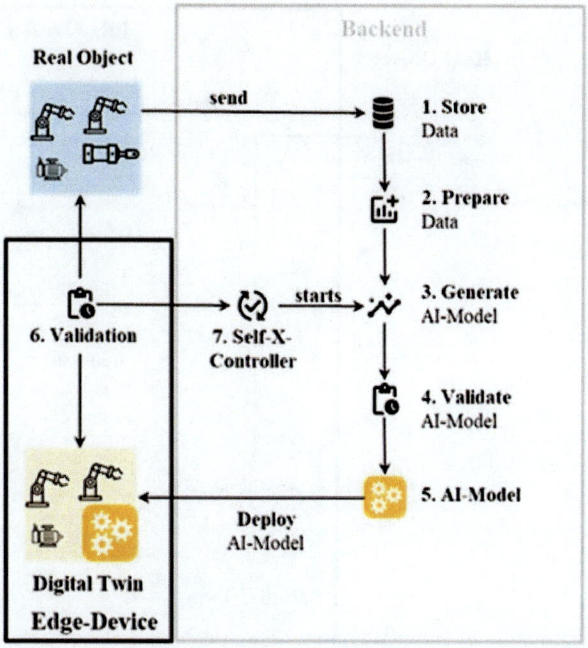

Fig. 11.3 Steps from the data with the Self-X-Controller

new model is sufficiently high, the new AI model can be deployed on the Edge Device at the machine [47].

The Self-X-Controller itself uses AI methods like machine learning in order to realize the Self-X properties for improving the production system. In this case, the backend generates the AI models with supervised self-learning systems. The AI models are deployed to the Edge Device at the machine by the backend. Again, if the accuracy is too low, the backend updates the AI model on the Edge Device (Fig. 11.3) [48].

At first, the real object sends the data to the backend. All received data will be stored (1). Step (2) consists of querying the data with defined rules to generate the AI model (3). The generated AI models are validated (4) with the data. If everything is sufficient, the new AI model (5) is complete and the backend deploys the AI model to the Edge Device. When the first AI model is deployed, the continuous validation process (6) between real and predict values starts. Should the validation process fail, a signal is sent to the Self-X-Controller. The Self-X-Controller (7) analyses the problem and starts the training (3) incorporating also all newly stored data. After the training finishes, the backend validates (4) the new AI model. If the accuracy of the new AI model (5) is improved, the backend deploys the new AI model to the Edge Device.

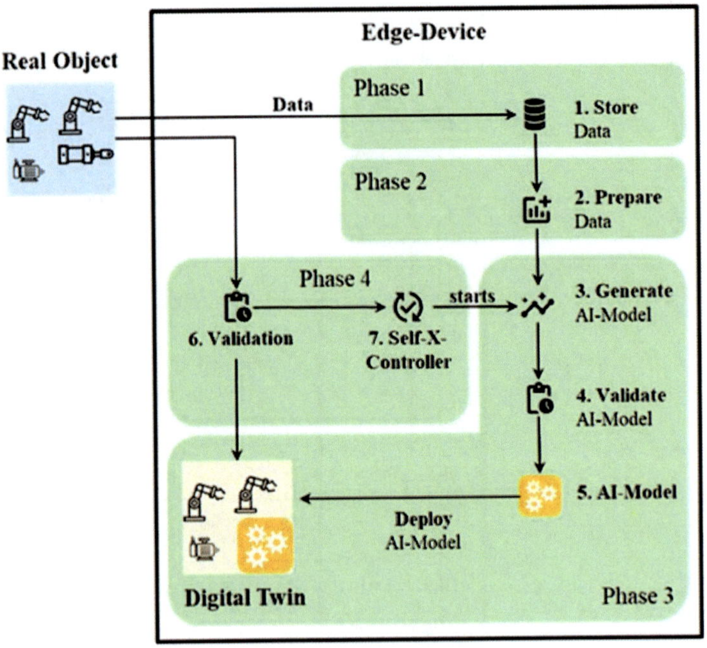

Fig. 11.4 Steps and phases on the Edge Device

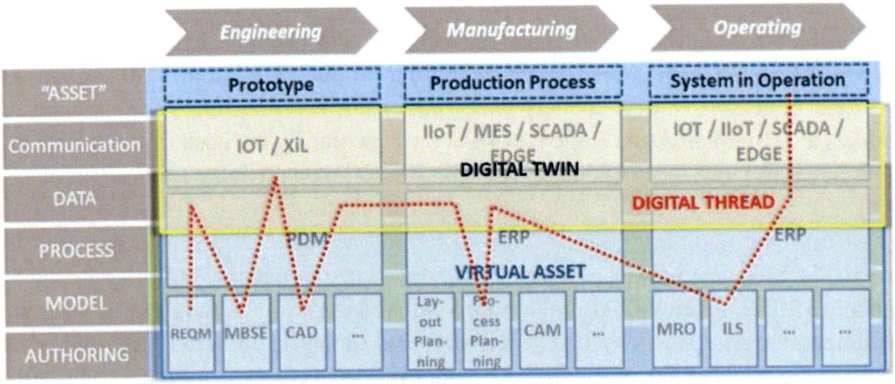

Fig. 11.5 Digital Thread connects lifecycle phases across six layers

11.4 Self-X Digital Twin on Edge-Devices

In the near future, the generation and roll-out of Digital Twins for production systems including AI-services is expected not to require a high degree of expertise in the field of AI-technologies. After the backend-supervised Self-X Digital Twins described in the previous chapter, this next step will help enabling the wide-spread usage of Digital Twins in production.

In this future step, a user gets a new Edge Device including its Digital Twin. This device has some interfaces like Profinet, Ethernet, etc. When the customer plugs the device into the switch cabinet, it configures the communication interface and establishes the kind of Digital Twin to be used [49].

Phase 1 storing and interpretation of all received signals. The device has no prior knowledge of the related physical processes. Between all signals, the most relevant ones must be determined and their types classified in the initial stage. In this phase, it is important to recognize underlying correlations and perform dimension reductions.

Phase 2 consists of the signal classification. After a certain time, from the incoming signals patterns are deduced like "the machine is in operation" or "the machine does not operate". The system analyses the data in more and more detail. After this, the system has classified the signals and recognized allowed and not allowed modes of operation for the machine.

Phase 3 is the model creation. The device uses all stored classifications and generates the initial AI model. It automatically calculates a topology of an AI model based on the selected signal classifications and trains the model using the associated data.

Phase 4 includes the continuous monitoring and self-optimization based on real-time data. The Digital Twin monitors the physical system by itself and optimizes processes on the basis of new data records.

When the device reaches the fourth and final phase, the device is able to modify itself for the first time allowing for self-optimisation. To ensure the accuracy of the used AI models, the Self-X-Controller begins to continuously monitor the difference between real and predicted values. If the deviation is too high, the Self-X-Controller triggers the learning process of the AI model, thus reducing the gap between real and predicted values. Once the gap is under a defined threshold, the Self-X-Controller stops the learning process. The device is also capable of learning and evaluating new AI models simultaneously to the ongoing operations. If a new AI model shows improved performance, that is a better approximation of the underlying incoming sensor data, the Self-X-Controller switches models and activates the improved AI model [50]. Figure 11.4 shows the assignment of the steps to the phases.

The process is analogous to backend-enables approach shown before, but now all functionality is located on the Edge Device directly at the machine.

11.5 Trends in Digital Twin

In this section, trends in research and practice are identified. This overview covers different levels of system scope and technological uncertainty, as applied to the eight dimensions identified in the introduction. Together, this allows a more accurate expression of the eight dimensions in terms of aggregation levels [2].

Keeping the previous in mind, the following considers the trends evolving in the different dimension of the Digital Twin respectively. Although the taxonomy is not subdivided in further detail and all dimensions are equally weighted, this structural classification of dimensions is not as homogeneous as shown. Within the dimensions, the handling of data is a prominent topic. Looking at the recent literature, the future Digital Twin will have the following properties regarding data handling [51]: rapid bi-directional data links [52], comprehensive processing of data [53], and performant synchronization between the physical counterpart and Digital Twin (continuous or periodical) [54]. Building on a deep integration of Digital Twins, modern technologies like CPS and IoT play a crucial role, but the further development of the Digital Twin is not limited to their inclusion [55].

Furthermore, while a Digital Twin itself is a deeply integrated system, its integration with further IT systems is of crucial importance. The concept of Configuration Lifecycle Management (CLM) has a significant influence on the Digital Twin, which makes it possible to create temporally valid views of the product across IT system boundaries and to manage the product configurations across all phases of the product lifecycle. The essential function of CLM is to generate the various views of the digital product model in the course of the lifecycle, to keep them consistent and to document their validity over time. It uses cross-system and cross-disciplinary baselines for this purpose. These baselines document the status of the configuration at a specific point in time or the degree of maturity and, therefore, also control the representation of the Digital Twin. Baselines enable companies to immediately and reliably respond to the question of whether and how the product or asset meets the requirements placed on it at any point in the process or in what state the asset was at a defined point in time, for example which product configuration was delivered to the customer [56].

A high-performance PLM integration platform, containing connectors to all involved IT systems, is required to manage the configuration of a product in a traceable manner along its entire life cycle (Fig. 11.5). The higher the number of different systems interacting across various domains, the higher the importance of proper traceability of data and decisions. As an intermediate layer across IT systems, it creates the prerequisites for bringing together the information from the individual IT systems in a manner corresponding to the Digital Thread concept. Here, the Digital Thread is the unifying element which connects all phases of product lifecycle with layers of the enterprise architecture [56].

In industries such as mechanical and plant engineering or shipbuilding, enterprises face the challenge that the manufacturer who builds and provides the Digital Twin is not necessarily the operator and user who feeds it with operating data. Therefore, both the digital data and the operating data or at least part of it must be exchanged and synchronized across companies in order to keep the Digital Twin up-to-date and to be able to use the operating data for the continuous improvement of the physical asset. Therefore, issues such as data security [57], protection of intellectual property [58], and ownership of data [59] play a very central role when setting up and using a Digital Twin application [56].

Today, more and more customers are demanding that their suppliers deliver digital data and models to support Digital Twin applications along with physical assets. With the help of CLM, users can not only control the scope of information provided, but also the level of detail of the information and the formats in which it is provided.

They can be compiled largely automatically and made available to the customer as a data package, for example in 3D PDF format [59].

The basis for traceability and CLM is the lightweight integration of the domains involved in the development and their IT tools (data source) through the generation of a consistent data model. The data objects contained in this data model serve as placeholders for the results that typically occur during the product development (e.g. based on the V-model): deliverables or configuration items (CIs). With the help of predefined link types (trace links) with permitted start and end data objects, traceability requirements across domains and IT tools are implemented. The data model and link types can easily be expanded or configured [56]. In this way, Digital Twins can be embedded in PLM system landscape.

Digital Twin as the means to work with the high-fidelity representation of the real world is gaining attention also outside the manufacturing industry. A comprehensive representation of trends may also include research from healthcare [60], social sciences [61], and economy [62].

The Digital Twin initiative aims at revolutionizing healthcare for the benefit of citizens and society through the creation of Digital Twins—computer models of individuals that allow identification of the individually best therapy, prevention or health care, making unavoidable mistakes (such as ineffective treatment recommendations) in complex situations safely, cheaply, and quickly on computer models of reality rather than in reality. For that, the vast and ever-growing knowledge base on biological mechanisms as well as healthcare and research data generated from countless individuals was leveraged to generate a generic reference model. This can be individualized with molecular, analytical and other diagnostic data from an individual into a Digital Twin, a digital self that accompanies each person from birth onwards, adapting and reacting as humans do. Digital Twins are accurate computer models of the key biological processes within every individual that keep humans healthy or lead to disease. Using these twins, individually optimal therapies, preventive or lifestyle measures can be identified, without exposing individuals to unnecessary risk and the healthcare system to unnecessary costs [60].

A virtual society can be built by a collection of Digital Twins which are autonomous systems (agents) and their activities. The aim of such Digital Twins is to represent and strengthen human activities in virtual societies. Such Digital Twins need to be able to interact with each other using operations and, at least in part, will inevitably need to be able to interact autonomously like human beings. Digital Twins will also carry out simulations and other calculations and actively feed the results back to the real targets. Thus, they will bridge the gap between the real and virtual. Digital Twins can be viewed as part of CPS when viewed together with real human beings or objects that have received feedback. Additionally, there are Digital Twins generated by other Digital Twins with properties that do not exist (called derivative Digital Twins). Digital Twins are autonomous agents that hold data for people and objects internally, thus can form some kind of society and need the appropriate environment. By giving the operating environments of Digital Twins or derivative Digital Twins high level functionality or even linking them with external applications, it will be possible to create various virtual societies [61].

The Digital Twin of the economy provides an economic tool and platform which allows the dynamic-experimental development as well as objective-transparent evaluation of macroeconomic policies and is constructed on two consecutive methods: Firstly, the building plan of the underlying economic model is established by an economic architecture, that is an investigative approach to uncover hidden contexts for more transparency, and, secondly, the complex nature of the economy, as designed or created after that blueprint, is then dynamically as well as realistically simulated by agent-based modelling. The purpose of this tool and its platform, apart from the provision of reliable recommendations for political decision-making, is the promotion of interdisciplinary research to enhance the state of knowledge in economics and to facilitate the transformation to a much more economically or ecologically sustainable as well as socially fairer economy or society. Macroeconomists and politicians get an adequate instrument which can more realistically replicate the real-life economy and, thus, can provide better (unbiased) advice [62].

11.6 Challenges in Digital Twin

In this section, challenges in research and practice for the near future are identified. This section builds upon the trends identified in Sect. 11.5—whereas current initiatives were described in detail there, this section will focus on major assumptions, limitations and gaps in Digital Twin research and applications.

Some authors perceive the greatest limitation of the definition of the Digital Twin in the restriction to "physical products" in the real space. Subsequently, it would be nonsensical to call the concept a Digital Twin as long as the physical pendant is not yet manufactured or existing [63]. Moreover, the physical twin emerges as the realization of a digital model. However, the Digital Twin consists of several use cases and is more of a whole strategy than just a single instance. The combination of a not yet existing physical entity in the real space can be seen as a Digital Twin, as long as a physical product is emerging in the near future and the use case contributes to the overall twinning strategy [63].

The first challenge is the implementation of the Digital Twin. Without a unified definition and standard architecture, a Digital Twin implementation is still no subject of systematic consolidation, and most studies as well as commercial offerings discover partial process views. For the conception of Digital Twin, first methodical results are available [63], based on Digital Twin types explained in Sect. 3.3. Such procedures facilitate the agile approaches for implementation of Digital Twin [64], based on a comprehensive Product Lifecycle Management (PLM). One of the observations for this gap between theory and practice is the different interpretations of how, in fact, a Digital Twin can be applied [65].

Similarly, the managerial contributions could be the possibility to develop a process model to standardize the implementation process and selection of components of a Digital Twin [3]. To achieve this goal, several, user-driven industry initiatives were constituted [66]. The aim of such associations is to bring the parallel development strands together to form an industrial Digital Twin and to develop it as an open-source solution together with the member companies. At this time, no practitioners can compare existing Digital Twins or those on development with taxonomies from the literature and equip their Digital Twins with all necessary characteristics. One of the first tasks is the definition of a reference architecture for Digital Twin, similar to RAMI [67].

With regard to the eight dimensions of Digital Twin, some challenges are weighed higher depending on the domain the Digital Twin is being implemented [68]. These challenges are primarily technical and reflect mainly the data handling: acquisition, collecting and processing of high-dimensional data; time series, multi-modal, and multi-source data communication. In a large organization, collecting data from a considerably large number of IoT devices, collating it according to time frequencies and preprocessing it for input to machine learning is a challenging big data task [69]. In many user scenarios, proper handling of Big Data is the decisive criterion for the implementation of Digital Twin [70].

High-fidelity bi-directional synchronization is especially challenging for large-scale industries, requires resources and high-stream Industrial Internet-of-Things (IIoT) connection [71]. For a proper function, Digital Twin demands synchronization of data, model, and service (DMS). An IoT system constructed by this method looks like a "device-Digital-Twin-application". It requires the specific architecture of the IIoT: local collection devices (the edge) and cloud systems implant the unified DMS framework to form partial Digital Twins with different functions, which logically constitute the Digital Twin of the equipment which it isolates the access between business applications and machines and makes the system more secure [72]. Compared with the method of Digital Twin deployment only in the cloud, this concept strengthens the collaboration between local and cloud, more-over promotes the synchronization of virtual and real [71].

Filling the gaps between virtual and physical systems by physically bound Digital Twin can open new perspectives in Smart Manufacturing. In this case, Digital Twin represents manufacturing cells, simulate system behaviors, predict process faults, and adaptively control manipulated variables. Apart of highly-performant interfaces, the manufacturing cell demands Machine Learning approaches for the industrial control process which is able to acquire process knowledge, schedule manufacturing tasks, identify optimal actions, and demonstrate control robustness. The intelligent control algorithms are being trained and verified upfront before deployed to the physical world for implementation [73]. Managing high dimensional data, with the various other software used by an industry and combining these with expert deep learning skills and equipment is a tedious task [69]. New methods like continual learning and federated learning seem promising for the use of Digital Twin but require further research [74, 75].

Accuracy and data quality (see Chap. 7) are challenging the Digital Twin in many aspects: using partial accuracy cany improve the overall system performance [76]. Improving hydrodynamic model accuracy without compromising computational efficiency has always been of high interest for safe and cost-effective marine operations. With continuous development of sensor technology and computational capacity, an improved Digital Twin concept for vessel motion prediction can be realized based on an onboard online adaptive hydrodynamic model. This makes possible a practical approach for tuning of important vessel hydrodynamic model parameters based on simulated onboard sensor data of vessel motion response [76].

One of the almost unlimited fields for Digital Twin is the fast and predictive identification of misbehavior and damage of technical products, based on a simplified framework in the context of dynamical systems. Especially promising looks the integration of physics-based models with Machine Learning in order to investigate several damage scenarios. This approach uses an interpretable model (physics-based) to build a fast Digital Twin that will be connected to the physical twin to support real time engineering decisions. Different classifiers and different model parameters can be considered to achieve the best accuracy. The most important advance is this approach is the possibility of integrating physics-based models with machine learning for different scenarios [77].

Human factors affect the implementation and operation of the Digital Twin in multiple roles and ways. In the phase of the implementation of a Digital Twin, building a software for Digital Twin requires a development team of software engineers and subject matter experts to test the suitability of the software for the particular task. That gives this development the transdisciplinary character [78]. Moreover, simulation-based optimization provides faster and efficient solutions. Conventionally, solutions calculated by analytical models are fed to the simulation software manually (for example, hardware-in-the-loop) rather than optimization being done on the simulation software by use of mechanisms like reinforcement learning [69]. For larger systems, which include supply chain networks and logistics, etc., global implementation level is more desirable [69].

For user and operator of Digital Twin, handling and presenting a huge amount of collected data and information in a Digital Twin during operation in an intuitive manner without a mental overload remains a challenge. By integrating graphics, audios and real-world objects, Augmented Reality (AR) facilitates the users to visualize and interact with Digital Twin data at a new level. AR gives the opportunity to provide intuitive and continual visualization of the Digital Twin data. The challenge is to move parts of the Human–Machine-Interface (HMI) to a portable AR device (Microsoft HoloLens, Google Glass, or similar) in order to visualize the Digital Twin data of a dedicated process in a real environment (e.g. manufacturing, maintenance). Such an application supports the operator to monitor and control the technical system (e.g. machine tool) at the same time, but also enables to interact and manage the Digital Twin data simultaneously, which provides an intuitive and consistent HMI [79]. In a logistics use case, a concept integrating the Digital Twin and adaptive automation was developed. The usage of a digital representation with adapting autonomy allows combining the strength of humans and machines in order that the operator uses his cognitive advantage to provide specific support when the machine reaches its limits [80].

Finally, the security is one of prevailed constraints of each IT system, and, therefore, remains a challenge for the Digital Twin, too. With the Digital Twin operating across multiple industrial partners and inventory sites, the growing security concerns are inevitable. Not only the cross-industry ownership and security concerns, in particular to shared data, but also the leak of real-time monitoring data can be hazardous to an enterprise [69]. The preventing measures can be organizational [58] and technical [57]. While industrial environments are increasingly equipped with sensors and integrated to enterprise networks, attack surface grows continuously and demands new approaches. Otherwise, Digital Twins provide a chance for protection and can contribute to enterprise security by simulating attacks and analyzing the effect on the virtual counterpart. However, the integration of Digital Twin security simulations into enterprise security strategies is currently neglected. This task is provided by Security Operation Center (SOC) which takes the challenge to develop a process-based security framework to incorporate Digital Twin security simulations in the SOC [81]. However, the insufficient security remains as limitation of Digital Twin.

11.7 Closing Remarks and Conclusions

The rising digitalization is changing almost every segment of our daily lives in business and leisure. As just one expression of this, Digital Twins are as a powerful means to optimize the reality using its high-fidelity digital surrogate. Up to now, this huge potential was primarily recognized in the manufacturing industry but also in healthcare, society, and economics. In this chapter, trends, challenges and future perspectives associated with Digital Twin were presented. As it has become evident, the Digital Twin is an encompassing approach of digitization for tackling complex, real-world problems. The Digital Twin is also an important approach in overarching approaches like Concurrent Engineering, Systems Engineering and Transdisciplinary Engineering. In essence, multiple disciplines, multiple functional roles, and multiple stakeholders need to collaborate in the processes making up the engineering, healthcare, societal, or economic systems. Moreover, a lifecycle perspective supported by a holistic PLM approach is essential in achieving a solution that is both useful and usable in the context in which the challenging user scenario exists.

Product lifecycle is distributed almost everywhere across the globe: product and process development, production, and exploitation of a product rarely are fixed to a dedicated place [82]. This almost endless distribution causes high challenges to users, their organization, methods, and tools in term of communication, efficiency, and interoperability. The synchronous availability of the digital and the physical product alone brings interoperability into play as an indispensable prerequisite. It is enforced by dynamic interaction between technical and social characteristics of new product, process, and service development. Considering the human factor, e.g. human–machine interaction [83], a Digital Twin becomes part of a social-technical system. However, a human could be also considered as the subject of Digital Twin (e.g. in medicine, pharmacy, or medical engineering). This moves the scope of Digital Twin in the area of transdisciplinary engineering [84, 85].

Throughout this chapter, several aspects of a specific expression of Digital Twin have been described based on the previous chapters of this book. For all the aspects making up a system, various challenges have been identified: efficiency, accuracy, speed, etc. In all aspects, the main challenge lies in a seamless integration of methods and tools that are suitable to support the dynamic and evolving nature of the modern production system that need to be developed including the development system itself. The important fact often overlooked is that these innovative solutions are available only recently and are subject of rapid improvement. The market share of products and services which encompasses Digital Twin dramatically rises.

Coming from the astronautics and aerospace with typical long-lasting products, the Digital Twin is often mentioned in context of PLM which is a related system to Digital Twin. In case of a one-of-a-kind-product, the Digital Twin comprise the auxiliary data of the PLM: the fit is almost complete. In other cases (serial products or products with a high variance), PLM manages several Digital Twins which can be different from each other depending on type and the underlying product. Generally speaking, this is not a static representation in PLM because two systems would be linked throughout the entire lifecycle of the system. The virtual and real systems would be connected as the system went through the four phases of creation, production (manufacture), operation (sustainment/support), and disposal [86].

Digital Twin is a powerful tool with widespread, adaptable capabilities combining simulation, autonomy, agent-based modelling, Machine Learning, prototyping, decision making self-optimisation, and Big Data into one [87]. Theses sub-systems should be tailored and prioritised depending on demand and specific user scenarios. The advancement and research in these subsystems at times create a hindrance for the development of Digital Twin. As up-and-coming the Digital Twin technology is, there are several technical and domain-dependent challenges that still need to be addressed [69]. One of them is certainly the presentation and visualization of the Digital Twin, its components and the operational outcome for the users and the stakeholders. However, a seamless integration of IoT, Machine Learning, and data is the distinguishing feature of the powerful and efficient product.

From its initial focus on manufacturing on, Digital Twin is spreading into different areas (e.g. automotive, healthcare, building technology, economy, social sciences education, etc.). Digital Twin is also aimed to optimize the product design and, at the same time, the design processes (i.e. concept generation, material selection, design verification, and decision making). Digital Twin can effectively assist the concept generation and redesign based on the data from existing product Digital Twin. Associated with other emerging technologies (i.e. big data analysis, AR/VR, cloud, edge computing, etc.), Digital Twin can be used to analyze a mass data from the real environment along with the whole product lifecycle and improve the visibility of design for verification. Digital Twin can play a vital role to simplify the design processes by employing digital prototyping, testing simulation, and prediction, but it still has many potential applications in the product design stage like self-optimization of products in service [88].

Despite this progress and individual project-based efforts, significant implementation gaps exist in the field, which have caused delay in the widespread adoption of this Digital Twin. Major reasons for this delay are the lack of a universal Digital Twin reference framework, domain dependence, security concerns of shared data, reliance of Digital Twin on other technologies, and lack of Digital Twin performance metrics [69]. The lack of standardization is particularly significant because it is unlikely that a complex system will come from a single source in a challenging environment. As a tool which can have many sub-components spread across collaborators and industry partners, developing regulations and security mechanisms is imperative for widespread of adoption of Digital Twin to overcome the concerns regarding data sharing [69]. There should, however, be openness to the input of other disciplines, including their methods and tools needed to deal with their aspect of the overall problem. Selecting methods, applying them, as well as further developing these methods in the context of complex societal problems, cannot be the task of one discipline alone [28].

References

1. Verhagen WJC, Stjepandić J, Wognum N (2015) Challenges of CE. In: Stjepandić J, Wognum N, Verhagen WJC (eds) Concurrent engineering in the 21st century. Springer International Publishing, pp 807–833
2. Verhagen WJC, Stjepandić J, Wognum N (2019) Future perspectives in systems engineering. In: Stjepandić J, Wognum N, Verhagen WJC (eds) Systems engineering in research and industrial practice. Springer International Publishing, pp 403–420
3. van der Valk H, Haße H, Möller F, Arbter M, Henning JL, Otto B (2020) A Taxonomy of digital twins. In: Anderson B, Thatcher J, Meservy R (eds) Proceedings of 26th Americas conference on information systems, pp 1–10
4. Jiang Z, Guo Y, Wang Z (2021) Digital twin to improve the virtual-real integration of industrial IoT. J Ind Inf Integr 22:100196. https://doi.org/10.1016/j.jii.2020.100196
5. Hong Lim KY, Zheng P, Chen CH (2020) A state-of-the-art survey of digital twin: techniques, engineering product lifecycle management and business innovation perspectives. J Intell Manuf 31:1313–1337. https://doi.org/10.1007/s10845-019-01512-w
6. Fuller A, Fan Z, Day C, Barlow C (2020) Digital twin: enabling technologies, challenges and open research. IEEE Access, 8, Art. no. 9103025, pp 108952–108971. https://doi.org/10.1109/ACCESS.2020.2998358
7. Redelinghuys AJH, Basson AH, Kruger K (2019) A six-layer architecture for the digital twin: a manufacturing case study implementation. J Intell Manuf 2019. https://doi.org/10.1007/s10845-019-01516-6
8. Lu Y, Xu X (2019) Cloud-based manufacturing equipment and big data analytics to enable on-demand manufacturing services. Robot Comput Integr Manuf 57:92–102
9. El-khoury J, Berezovskyi A, Nyberg A (2019) An industrial evaluation of data access techniques for the interoperability of engineering software tools. J Ind Inf Integr 15(2019):58–68
10. Lu Y, Liu C, Wang KIK, Huang H, Xu X (2020) Digital twin-driven smart manufacturing connotation, reference model, applications and research issues. Robot Comput Integr Manuf 61:101837
11. Leng J, Liu Q, Ye S, Jing J, Wang Y, Zhang C, Zhang D, Chen X (2020) Digital twin-driven rapid reconfiguration of the automated manufacturing system via an open architecture model. Rob Comput-Integr Manuf 63:101895
12. Bicocchia N, Cabria G, Mandreolia F, Mecella M (2019) Dynamic digital factories for agile supply chains: an architectural approach. J Ind Inf Integr 15:111–121
13. Erkoyuncu JA, del Amo IF, Ariansyah D, Bulk D, Vrabic R, Roy R (2020) A design framework for adaptive digital twins. CIRP Ann 69(1):145–148

14. Pang Z, Xie B, Zhu H, Luvisotto M (2019) Location aided commissioning of building automation devices enabled by high accuracy indoor positioning. J Ind Inf Integr 15:258–263
15. Mohamad EB, Ibrahim MAB, Sukarma L, Rahman MAA, Shibghatullah ASB, Salleh MRB (2017) Improved decision making in lean manufacturing using simulation-based approach. Int J Agile Syst Manage 10(1):34–48
16. Zheng P, Sivabalan AS (2020) A generic tri-model-based approach for product-level digital twin development in a smart manufacturing environment. Robot Comput-Integr Manuf, vol 64, August 2020. Article number 101958. https://doi.org/10.1016/j.rcim.2020.101958
17. Ladj A, Wang Z, Meski O, Belkadi F, Ritou M, Da Cunha C (2020) A knowledge-based digital shadow for machining industry in a digital twin perspective. J Manuf Syst. https://doi.org/10.1016/j.jmsy.2020.07.018
18. Okita T, Kawabata T, Murayama H, Nishino N, Aichi M (2019) A new concept of digital twin of artifact systems synthesizing monitoring/inspections, physical/numerical models, and social system models. Procedia-CIRP 79:667–672. https://doi.org/10.1016/j.procir.2019.02.048
19. Guo H, Zhu Y, Zhang Y, Ren Y, Chen M, Zhang R (2021) A digital twin-based layout optimization method for discrete manufacturing workshop. Int J Adv Manuf Technol 112:1307–1318. https://doi.org/10.1007/s00170-020-06568-0
20. Zhong RY, Ge W (2018) Internet of things enabled manufacturing: a review. Int J Agile Syst Manage 11(2):126–154
21. Cheng J, Zhang H, Tao F, Juang CF (2020) DT-II-digital twin enhanced industrial internet reference framework towards smart manufacturing. Rob Comput Integr Manuf 62 April 2020, 101881
22. Biahmou A, Emmer C, Pfouga A, Stjepandić J (2016) Digital master as an enabler for industry 4.0. Adv Transdiscip Eng 4, pp 672–681
23. Kaur MJ, Mishra VP, Maheshwari P (2020) The convergence of digital twin, IoT, and machine learning: transforming data into action, internet of things, pp 3–17. https://doi.org/10.1007/978-3-030-18732-3_1
24. Wang Y, Wang S, Yang B, Zhu L, Liu F (2020) Big data driven hierarchical digital twin predictive remanufacturing paradigm: architecture, control mechanism, application scenario and benefits. J Clean Prod 248:119299
25. Schleich B, Dittrich MA, Clausmeyer T, Damgrave R, Erkoyuncu JA, Haefner B, Wuest T (2019) Shifting value stream patterns along the product lifecycle with digital twins. Procedia CIRP 86:3–11
26. Adamenko D, Kunnen S, Pluhnau R, Loibl A, Nagarajah A (2020) Review and comparison of the methods of designing the Digital Twin. Procedia CIRP 91:27–32
27. Ying L, Yongping R, Jiamin W, Chao-Hsiu L (2020) BIM based cyber-physical systems for intelligent disaster prevention. J Ind Inf Integr 20:100171
28. Biesinger F, Meike D, Kraß B, Weyrich M (2018) A case study for a digital twin of body-in-white production systems—general concept for automated updating of planning projects in the digital factory. In: 23rd international conference on emerging technologies and factory automation (ETFA). IEEE. 04–07 September 2018, pp 19–26
29. Lu Q, Chen L, Li S, Pitta M (2020) Semi-automatic geometric digital twinning for existing buildings based on images and CAD drawings. Autom Constr 115:103183
30. Sieg O, Völl C (2020) Digital twin and Product Lifecycle Management (PLM) as part of the corporate strategy [Digital Twin und PLM als Teil der Unternehmensstrategie], ZWF Zeitschrift fuer Wirtschaftlichen Fabrikbetrieb, vol 115, April 2020, p 38
31. Booyse W, Wilke DN, Heyns S (2020) Deep digital twins for detection, diagnostics and prognostics. Mech Syst Signal Process 140:106612
32. Gürdür Broo D, Boman U, Törngren M (2021) Cyber-physical systems research and education in 2030: scenarios and strategies. J Ind Inf Integr 21:100192
33. Martínez GS, Sierla SA, Karhela TA, Lappalainen J, Vyatkin V (2018) Automatic generation of a high-fidelity dynamic thermal-hydraulic process simulation model from a 3D plant model. IEEE Access 6:45217–45232

34. Wognum N, Mo J, Stjepandić J (2019) Transdisciplinary engineering systems. In: Kenett RS, Swarz RS, Zonnenshain A (eds) Systems engineering in the fourth industrial revolution: big data, novel technologies, and modern systems engineering, John Wiley & Co., Hoboken, pp 483–510
35. Avazpour I, Grundy J, Zhu L (2019) Engineering complex data integration, harmonization and visualization systems. J Ind Inf Integr 16:100103
36. Pfouga A, Stjepandić J (2018) Leveraging 3D geometric knowledge in the product lifecycle based on industrial standards. J Comput Des Eng 5(1):54–67
37. Chen Y (2020) A Survey on industrial information integration 2016–2019. J Ind Integr Manage 5(1):33–163. https://doi.org/10.1142/S2424862219500167
38. Xu LD (2020) Industrial information integration—an emerging subject in industrialization and informatization process. J Ind Inf Integr 17:100128
39. Donoghue I, Hannola L, Papinniemi J, Mikkola A (2018) The Benefits and impact of digital twins in product development phase of PLM. In: Chiabert P, Bouras A, Noël F, Ríos J (eds) PLM 2018: product lifecycle management to support industry 4.0. Springer International Switzerland, pp 432–441
40. Riascos Castaneda R, Ostrosi E, Majić T, Stjepandić J, Sagot JC (2020) A method to explore product risk in product lifecycle management of configured products. In: Proceedings of the design society: DESIGN conference, pp 687–696
41. Denkena B, Doreth K, Köller M, Wilmsmeier S, Winter F (2020) Simulationsbasierte kombinierte Instandhaltungs- und Produktionsplanung, Ablaufsimulation in der Automobilindustrie. Springer Vieweg, Hrsg G, Mayer C, Pöge S, Spieckermann S, Wenzel S, pp 261–273
42. Donhauser T, Endter U, Schmidt T, Baier L, Schuderer P, Franke J (2020) Echtzeitabbild der Produktion. wt werkstattstechnik 110(4):214–219
43. Meudt T, Pohl M, Metternich J (2017). Die Automatisierungspyramide-Ein Literaturüberblick, TU Darmstadt, Darmstadt
44. Dehraj P, Sharma A (2021) A review on architecture and models for autonomic software systems. J Supercomput 77:388–417. https://doi.org/10.1007/s11227-020-03268-0
45. Tomforde S, Rudolph S, Bellman K, Würt RP (2016) An organic computing perspective on self-improving system interweaving at runtime. In: 2016 IEEE international conference on autonomic computing (ICAC), Würzburg, pp 276–284. https://doi.org/10.1109/ICAC.2016.15
46. Tahir M, Mamoon AQ, Dabbagh M, Computing TEA (2019) In: IoT ecosystem IEEE intl conf on dependable, autonomic and secure computing, intl conf on pervasive intelligence and computing, intl conf on cloud and big data computing, intl conf on cyber science and technology congress (DASC/PiCom/CBDCom/CyberSciTech). Fukuoka, Japan 2019:646–651. https://doi.org/10.1109/DASC/PiCom/CBDCom/CyberSciTech.2019.00122.)
47. Kritzinger W, Traar G, Sihn W (2018) Digital twin in manufacturing: a categorical literature review and classification. In: IFAC-Papers OnLine, vol 51, no 11, pp 1016–1022
48. Zhou G, Zhang C, Li Z, Ding K, Wang C (2020) Knowledge-driven digital twin manufacturing cell towards intelligent manufacturing. Int J Prod Res 58(4):1034–1051
49. Ramakrishnan R, Pendse A, Sharma C, Chimurkar P (2020) Speed breaker detection and mapping using IoT. In: Proceedings of the 3rd international conference on smart systems and inventive technology, ICSSIT 2020, 9214224, pp 294–299
50. Tomforde S, Gruhl C (2020) Fairness, performance, and robustness: is there a CAP theorem for self-adaptive and self-organising systems? In: Proceedings—2020 IEEE international conference on autonomic computing and self-organizing systems companion, ACSOS-C 2020, 9196252, pp 54–59
51. Sanchez M, Exposito E, Aguilar J (2020) Autonomic computing in manufacturing process coordination in industry 4.0 context. J Ind Inf Integr 19:100159
52. Ruppert T, Abonyi J (2020) Integration of real-time locating systems into digital twins. J Ind Inf Integr 20:100174
53. Wallis R, Stjepandic J, Rulhoff S, Stromberger F, Deuse J (2014) Intelligent utilization of digital manufacturing data in modern product emergence processes. Adv Transdiscipl Eng 261–270. https://doi.org/10.3233/978-1-61499-440-4-261

54. Perno M, Hvam L (2020) Developing a framework for scoping digital twins in the process manufacturing industry. Adv Transdiscipl Eng 13:475–486
55. Zhang C, Chen Y (2020) A review of research relevant to the emerging industry trends: industry 4.0, IoT, blockchain, and business analytics. J Ind Integr Manage 5(1):165–180
56. Wagner L (2020) Kein digital twin ohne digitale Durchgängigkeit. Zeitschrift für den wirtschaftlichen Fabrikbetrieb, Band 115, April 2020, pp 43–45. https://doi.org/10.3139/104.112315
57. Stjepandić J, Liese H, Trappey AC (2015) Intellectual Property Protection. In: Stjepandić J et al (eds) Concurrent engineering in the 21st century: foundations. Developments and challenges. Springer International Publishing, Cham, pp 521–552
58. Liese H, Stjepandic J, Rulhoff S (2010) Securing product know-how by embedding IP-protection into the organization. In: 2010 IEEE international technology management conference. ICE 2010:7477025
59. Kim JH (2017) A survey of IoT security: risks, requirements, trends, and key technologies. J Ind Integr Manage 2:1750008
60. DigiTwins NN (2021) Digital twins for better health. https://www.digitwins.org/. Accessed 30 Jan 2021
61. Takahashi K (2020) Social issues with digital twin computing. NTT Tech Rev. https://www.ntt-review.jp/archive/ntttechnical.php?contents=ntr202009fa5.html. Accessed 30 Jan 2021
62. Pobuda P (2020) The digital twin of the economy: proposed tool for policy design and evaluation. Real-World Econ Rev 94:140–148. http://www.paecon.net/PAEReview/issue94/Pobuda94.pdf. Accessed 30 Jan 2021
63. Trauer J, Schweigert-Recksiek S, Engel C, Spreitzer K, Zimmermann M (2020) What is a digital twin?—definitions and insights from an industrial case study in technical product development. In: International design conference—design 2020. https://doi.org/10.1017/dsd.2020.15.
64. Riascos R, Ostrosi E, Sagot J-C, Stjepandic J (2020) An approach for agile engineering change management within global product development. Adv Transdiscipl Eng 12:584–593
65. Jones D, Snider C, Nassehi A, Yon J, Hicks B (2020) Characterising the digital twin: a systematic literature review. CIRP J Manuf Sci Technol 29(Part A):36–52. https://doi.org/10.1016/j.cirpj.2020.02.002
66. Nutzerorganisation NN (2021) Industrial digital twin association gegründet. https://www.vdma.org/v2viewer/-/v2article/render/52443301. Accessed 20 Jan 2021
67. DIN SPEC 91345 (2016) Referenzarchitekturmodell Industrie 4.0 (RAMI4.0), Deutsches Institut für Normung (DIN) e.V. https://doi.org/10.31030/2436156
68. Kutzke DT, Carter JB, Hartman BT (2021) Subsystem selection for digital twin development: a case study on an unmanned underwater vehicle. Ocean Eng 223:108629
69. Sharma A, Kosasih E, Zhang J, Alexandra A, Calinescu A (2020) Digital twins: state of the art theory and practice, challenges, and open research questions. arXiv:2011.02833v3
70. De Modesti PH, Carvalhar Fernandes E, Borsato M (2020) Production planning and scheduling using machine learning and data science processes. Adv Transdiscipl Eng 13:155–166
71. Zhang K, Qu T, Zhou D, Jiang H, Lin Y, Li P, Guo H, Liu Y, Li C, Huang GQ (2020) Digital twin-based opti-state control method for a synchronized production operation system. Robot Comput Integr Manuf 63:101892
72. Biahmou A, Stjepandic J (2016) Towards agile enterprise rights management in engineering collaboration. Int J Agile Syst Manage 9:302–325
73. Xia K, Sacco C, Kirkpatrick M, Saidy C, Nguyen L, Kircaliali A, Harik R (2020) A digital twin to train deep reinforcement learning agent for smart manufacturing plants: environment, interfaces and intelligence. J Manuf Syst (in press)
74. Li D, Liu S, Gao F, Sun X (2021) Continual learning classification method with constant-sized memory cells based on the artificial immune system. Knowl-Based Syst, 213:106673
75. Zhang W, Li X, Ma H, Luo Z, Li X (2021) Federated learning for machinery fault diagnosis with dynamic validation and self-supervision. Knowl-Based Syst 213:106679
76. Han X, Leira BJ, Sævik S (2021) Vessel hydrodynamic model tuning by discrete Bayesian updating using simulated onboard sensor data. Ocean Eng 220:108407

77. Ritto TG, Rochinha FA (2021) Digital twin, physics-based model, and machine learning applied to damage detection in structures. Mech Syst Signal Process 155:107614
78. Mo JPT, Beckett RC (2019) Architectural modeling of transdisciplinary system with inherent social perspectives. J Ind Integr Manage 1950012. https://doi.org/10.1142/S242486221950012X
79. Zhu Z, Liu C, Xu X (2019) Visualisation of the digital twin data in manufacturing by using augmented reality. Procedia CIRP 81:898–903
80. Wilhelm J, Beinke T, Freitag M (2020) Improving Human-machine interaction with a digital twin. In: Freitag M, Haasis HD, Kotzab H, Pannek J (eds) Dynamics in logistics. LDIC 2020. Lecture notes in logistics. Springer, Cham. https://doi.org/10.1007/978-3-030-44783-0_49
81. Dietz, M, Vielberth, M, Pernul, G. (2020) Integrating digital twin security simulations in the security operations center. In: ACM international conference proceeding series 25 August 2020 15th international conference on availability, reliability and security, ARES 2020; Virtual, Online; Ireland; 25 August 2020 through 28 August 2020; Code 162046
82. Voell C, Chatterjee P, Rauch A, Golovatchev J (2018) How digital twins enable the next level of PLM—a guide for the concept and the implementation in the internet of everything era. In: Chiabert P et al (eds) Product lifecycle management to support industry 4.0. Springer International, Cham, pp 238–249
83. Ma X, Tao F, Zhang M, Wang T, Zuo Y (2019) Digital twin enhanced human-machine interaction in product lifecycle. Procedia CIRP 83:789–793
84. Wognum N, Bil C, Elgh F, Peruzzini M, Stjepandić J, Verhagen WJC (2019) Transdisciplinary systems engineering: implications, challenges and research agenda. Int J Agile Syst Manage 12(1):58–89
85. Wognum N, Bil C, Elgh F, Peruzzini M, Stjepandić J, Verhagen W (2018) Transdisciplinary engineering research challenges. Adv Transdiscipl Eng 7:753–762
86. Grieves M, Vickers J (2017) Digital twin: mitigating unpredictable, undesirable emergent behavior in complex systems. In: Kahlen FJ, Flumerfelt S, Alves A (eds) Transdisciplinary perspectives on complex systems: new findings and approaches. Springer International, Cham, pp 85–113
87. Zhao C, Chen M, Wang L, Zhang L (2020) Collaboration of smart device in cloud manufacturing: a case of active recommendation model based on service agent. Adv Transdiscipl Eng 13:551–562
88. Lo CK, Chen CH, Zhong RY (2021) A review of digital twin in product design and development. Adv Eng Inf (in press)